BESTSELLING BOOK SERIES

The Origins of the U... For Dummies®

A Rough Timeline of the Universe

Probing back to the origin of the universe involves a lot of estimation and guesswork. Imagine that the word 'roughly' is written before each date!

- **0: The Big Bang.** Time and space are created. (Chapter 6 has more about the Big Bang.)
- **10^{-43} seconds: Gravity separates from the three other fundamental forces (electromagnetism, and the weak and strong nuclear forces).** This is the earliest time that theoretical physicists have probed so far. The strong force and electroweak forces become distinct soon after. (We explain electromagnetism in Chapter 4.)
- **10^{-35} seconds: The universe expands, undergoing a spectacular acceleration known as inflation.** This process takes tiny regions of space and blows them up into much larger volumes, ironing out any wrinkles in the process.
- **10^{-6} seconds: Particles gain mass.** The electroweak forces break down into the electromagnetism and weak forces we observe today. Sub-atomic particles gain mass.
- **1 second: The first composite particles.** Protons and neutrons form from a very hot soup of quarks and gluons (Chapter 9 explains this in more detail).
- **3 minutes: The first elements (mostly hydrogen and helium) form.** The universe expands and cools so fast that heavier elements don't even have a chance to be created.
- **380,000 years: The universe's temperature drops enough for the protons and neutrons to begin capturing electrons.** Also, for the first time, light travels freely through space, and the fog of the early universe clears. This light is still detectable today as the *cosmic microwave background* (head to Chapter 6 for an explanation of the CMB).
- **30 million years: Stars first appear in the universe.** Computer models suggest that the first stars may have formed at this point, along with the creation of heavy elements.
- **200 million years: The Milky Way forms.** In 2004, scientists calculated that the Earth's home galaxy was formed not long (well, in cosmic terms) after the first stars. Turn to Chapter 13 to find out about the creation of solar systems.
- **9 billion years: The Earth's solar system forms.** The disk of material left over after the formation of the Sun begins to get clumpy.
- **10 billion years: Life on Earth begins.** The effect of harsh solar radiation and lightning on a primordial soup of organic material may have kickstarted life.
- **11 billion years: Oxygen accumulates in the atmosphere of the Earth.** The essential gas for animals to breathe appears for the first time.
- **13.5 billion years: Early humans evolve in Africa.** Modern humans first appear in the continent and colonise the rest of the planet.
- **13.7 billion years: You pick up this book.** The origins of the universe are explained!

For Dummies: Bestselling Book Series for Beginners

The Origins of the Universe For Dummies®

Glossary

Anisotropy: The variation of a physical property, depending on the direction in which it's measured. For example, the temperature of the cosmic microwave background radiation is anisotropic.

Antimatter: Material composed entirely of antiparticles.

Antiparticle: A counterpart to every particle having the same mass but opposite properties at a quantum level (electrical charge, for example). The positively charged positron is the antiparticle of the negatively charged electron, for example.

Baryon: The family of heavy subatomic particles, which includes protons and neutrons, that interact through the strong nuclear force.

Big Bang: The widely accepted theory that says the universe started expanding roughly 14 billion years ago from an extremely dense and hot state.

Black hole: An object with such a strong gravitational field that nothing inside it can escape, including light.

Boson: Force-carrying fundamental particles, such as the photon (electromagnetic force) or the W and Z particles (weak force).

Cosmic microwave background (CMB): The cooled remnant of the Big Bang, this microwave radiation fills the entire universe and can be observed today with an average temperature of about 2.725 kelvin.

Dark energy: A mysterious energy, thought to make up 70 per cent of the universe, which causes the universe to expand more and more quickly.

Dark matter: Unknown substances that are detectable in space by their gravitational effects, but which don't shine like normal matter.

Doppler effect: The process by which the frequency or wavelength of light or sound seems to be altered by the motion of its source relative to the observer.

Electron: A light fundamental particle with a negative charge.

Fundamental particle: A particle, such as the quark or electron, which scientists believe cannot be subdivided further.

Galaxy: An enormous system containing billions of stars, plus vast amounts of dust and gas; the Milky Way, for example.

Hubble constant: The ratio of the speed with which galaxies are moving away (receding) from an observer to their distance from us, due to the expansion of the universe.

Nebula: A cloud of gas and dust in space that may emit, reflect, and/or absorb light.

Neutrino: A fundamental particle that has no electric charge and very little mass. It can pass through whole planets or stars without interacting with other particles.

Neutron: One of the baryons that make up atoms. Neutrons have no electrical charge and are made of one up quark and two down quarks.

Neutron star: The collapsed core of a massive star that remains after a supernova explosion. The remaining matter is compressed so tightly that negatively charged electrons and positively charged protons are forced together.

Photon: A small packet of electromagnetic radiation.

Planet: A large, near-spherical object that orbits a star.

Proton: A baryon found in the nucleus of every atom. Made of two up quarks and one down quark, a proton has a positive charge.

Pulsar: A fast-spinning, dense neutron star that emits light, radio waves, and/or X-rays in beams like the light from a lighthouse.

Quantum: The smallest possible unit of something that can exist.

Quark: The family of fundamental particles that combine to make baryons. Quarks come in six flavours: up, down, charm, strange, top, and bottom.

Quasar: The bright centre of an active galaxy, probably fuelled by an enormous black hole that swallows matter.

Red giant: A large, bright star with a low surface temperature.

Red-shift: An increase in the wavelength of light or sound. In the case of distant galaxies, this increase is caused by the expansion of the universe.

SETI: The search for extraterrestrial intelligence.

Solar system: The Sun and all the celestial bodies that orbit it, including the planets and their moons, asteroids, comets, and so on.

Star: A large mass of hot gas held together by its own gravity and fuelled by nuclear reactions.

String theory: Theory of the universe, which says that the fundamental ingredients of nature are tiny, one-dimensional filaments called strings.

White dwarf: The remnant core of a star that has exhausted its nuclear fuel and settled into a solid ball of matter.

The Origins of the Universe FOR DUMMIES®

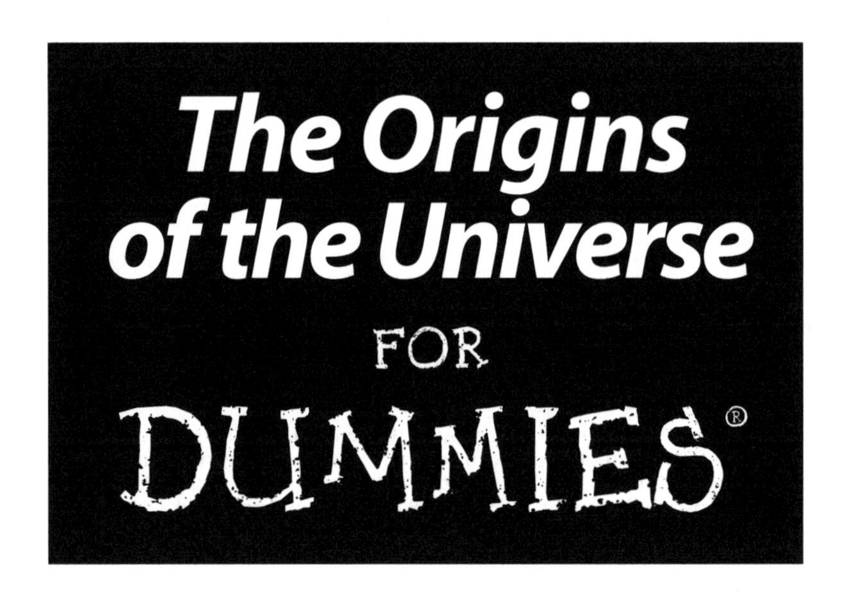

The Origins of the Universe For Dummies®

by Stephen Pincock and Mark Frary

John Wiley & Sons, Ltd

The Origins of the Universe For Dummies®

Published by
John Wiley & Sons, Ltd
The Atrium
Southern Gate
Chichester
West Sussex
PO19 8SQ
England

E-mail (for orders and customer service enquires): cs-books@wiley.co.uk

Visit our Home Page on www.wiley.com

Published by John Wiley & Sons, Ltd, Chichester, West Sussex

British Library Cataloguing in Publication Data: A catalogue record for this book is available from the British Library.

ISBN: 978-0-470-51606-5

Printed and bound in Great Britain by Bell and Bain Ltd, Glasgow

10 9 8 7 6 5 4 3 2 1

About the Authors

Stephen Pincock has been writing about science for the past 15 years, after finishing a degree in Microbiology at the University of New South Wales, Australia, and realising that while the whole science thing is utterly fascinating, he was less than eager to spend the rest of his life peering down a microscope.

Stephen's currently a regular science contributor to *The Financial Times* and *The Lancet* among many other publications, and is the international correspondent for *The Scientist*. For quite a while he was an editor at Reuters Health.

Mark Frary is a science and technology writer. He studied astronomy and physics at University College London, writing a dissertation on the production of positronium. While there, he worked at the Mullard Space Science Laboratory on atmospheric plasma physics. After completing his degree, he moved to Geneva and worked on the OPAL experiment at the European particle physics laboratory CERN.

Mark co-wrote the book *You Call This The Future?*, a look at the 50 best science-fiction gadgets ever conceived and how they have become reality. He lives in Ampthill in Bedfordshire with his wife and two children.

Mark and Stephen are the authors of *Codebreaker: The History of Secret Communication*.

Dedication

We would like to dedicate this book to our long-suffering families: to Amanda, Daniel, Emily, Clare, Lola, and Seth with all our love.

Authors' Acknowledgements

From Stephen: Throughout the writing of this book I've relied on the vital support of my lovely family, and I'd like to thank them all for their forbearance during those periods when, instead of being with them, I seemed to do little other than write.

I'm also very glad to have worked with the Wiley team on this project, including Rachael, whose encouragement was sometimes the only thing that got me through the work. Thanks also to the whole editing team, particularly Brian and his remarkable ability to carve lucid sentences from sometimes unpromising raw material!

Finally, thanks to Mark, whose talent and knowledge seemingly know no bounds.

From Mark: Thanks to my family and friends – in particular Clare, Margaret, Graham, mum, and dad – for giving me the time to write this book and for not minding too much about the late evenings and weekends working which should have been spent with them. Thanks also to my co-author Stephen Pincock, who has provided unstinting support and given me a huge number of welcome suggestions that have improved the book considerably.

I'm grateful to Jason Dunne, Rachael Chilvers, and the rest of the team at Wiley for giving this project the green light and for their encouragement and suggestions throughout the publishing process.

A big thanks also to Dr Carolin Crawford for her input on the technical aspects of this book.

Publisher's Acknowledgements

We're proud of this book; please send us your comments through our Dummies online registration form located at www.dummies.com/register/.

Some of the people who helped bring this book to market include the following:

Acquisitions, Editorial, and Media Development

Project Editor: Rachael Chilvers

Development Editor: Brian Kramer

Content Editor: Nicole Burnett

Copy Editor: Andy Finch

Proofreader: Charlie Wilson

Technical Editor: Dr Carolin Crawford

Executive Editor: Jason Dunne

Executive Project Editor: Martin Tribe

Cover Photo: © Chris Collins/CORBIS

Cartoons: Ed McLachlan

Special Help: Zoë Wykes

Composition Services

Project Coordinator: Erin Smith

Layout and Graphics: Reuben W. Davis, Brooke Graczyk, Joyce Haughey, Melissa K. Jester

Proofreader: Laura Albert

Indexer: Cheryl Duksta

Wiley Bicentennial Logo: Richard J. Pacifico

Publishing and Editorial for Consumer Dummies

Diane Graves Steele, Vice President and Publisher, Consumer Dummies

Joyce Pepple, Acquisitions Director, Consumer Dummies

Kristin A. Cocks, Product Development Director, Consumer Dummies

Michael Spring, Vice President and Publisher, Travel

Kelly Regan, Editorial Director, Travel

Publishing for Technology Dummies

Andy Cummings, Vice President and Publisher, Dummies Technology/General User

Composition Services

Gerry Fahey, Vice President of Production Services

Debbie Stailey, Director of Composition Services

Contents at a Glance

Table of Contents

Introduction

The book you hold in your hands has at its heart perhaps the most fundamental question anyone can ask him or herself: Where did I come from? (No, not the town you were born in!)

With *The Origins of the Universe For Dummies*, we want to do nothing less than take you on a tour back in time – through close to 14 billion years of history – to the point at which the vast majority of scientists believe the universe began. Don't worry, you don't need to be Stephen Hawking to understand what we'll be talking about.

Yes, going back that far in time is tough when most people can't remember what they ate for lunch last Wednesday. But stout-hearted scientists are trying to put together a complex jigsaw puzzle of pieces that stretch to far beyond the birth of humanity, beyond the birth of the Earth, beyond even the creation of the Earth's Sun and the galaxy that you inhabit, the Milky Way.

Care to join us? All you need for the journey is an open mind and a sense of wonder and questioning. You can leave your calculator at home. You won't be needing it.

About This Book

A multitude of theories exist for how the universe began. Some of them are scientific, some religious, some cultural, and some just plain crazy. What we try to do in this book is focus on the first of these categories.

Yet even within science a large number of different ideas are competing to explain how everything began. As a result, we concentrate on the most widely accepted scientific theory – an explanation known as the Big Bang.

The experiments that scientists are conducting to explore the early days of the universe are some of the most complex to get your head around. That's why, in this book, we explain why scientists believe in the Big Bang. We also come up with some everyday ideas that help you to understand what happened billions and billions of years ago.

The topics we address in the book are certainly challenging – why space is considered curved, how time is not the same for everyone, and the fact that everything you see is made up of strange objects such as quarks and leptons. These concepts and others may sound hard to grasp, but we believe that we've cracked them. Never again are you going to be stumped by a pub quiz question about how Hubble's Law affects the birth of the universe! Along the way, you also discover why the night sky isn't ablaze with light, what black holes really are, and how cosmologists have cleverly come up with all sorts of experiments to explain what really happened all those years ago.

Conventions Used in This Book

To help you grasp the universe while grasping this book in your hands, we have used the following conventions:

- *Italic* text highlights new words and defined items.
- **Bold** text indicates keywords in bulleted lists.
- `Monofont` text indicates a Web address.

What You're Not to Read

This book is not meant to be some boring lecture about astrophysics. It's meant to be informative and fun. Feel free to dip in and out at your leisure because each chapter and section is intended to be self-contained.

Within most chapters, you also find some additional information presented in shaded boxes. Although information in these boxes helps to give you the big picture, reading them isn't an essential part of understanding this book's main concepts.

As you read the book, you also encounter sections marked with the Technical Stuff icon. If you're not a fan of equations, you may want to skip over these bits. You can still grasp the book's main concepts without consuming these in-depth parts.

Foolish Assumptions

We don't assume much really, except that you have a curiosity about the world and universe around you. A little knowledge of mathematical equations is handy at times – the Appendix explains equations that may be helpful if you haven't been in a classroom for a few years – but you don't need to be a boffin to enjoy this book. Our view is that if we can't explain something complex in simple terms, it's not worth knowing.

How This Book Is Organised

The table of contents tells you the intricate details about how the parts, chapters, and sections of this book are organised, but here's a handy rundown of the major parts into which we divide the book.

Part I: In the Beginning: Early Ideas About Our Universe

We start the book by considering how the ancients viewed what was going on in the heavens and how religion, culture, and spirituality were the guiding principles in their explanations of the nature of the Sun, the Moon, and the stars.

The first scientific explanation of the universe came with the rise of the ancient Greeks. We examine how Greek scholars first postulated that the Moon shines because it reflects sunlight and how they began to predict lunar and solar eclipses by realising that many heavenly bodies travel in regular orbits.

We then discuss the cosmological revolution that began with Copernicus and his idea that the Earth isn't the centre of everything. This radical notion was the first small step in realising that the universe is an enormous place and that the Earth doesn't have a particularly special place within it.

Two centuries later, the life and works of Sir Isaac Newton – certainly one of the most influential of the pre-modern era cosmologists – shaped human understanding of the universe with his laws of motion and his understanding of gravity.

Part II: Modern Cosmology: Going Off with a Bang

The current understanding of the origins of the universe dates from the beginning of the 20th century. In this part, you meet Albert Einstein and explore his special and general theories of relativity. You also find out why things don't travel faster than the speed of light, what $E = mc^2$ really means, and what Einstein believed was his greatest blunder.

Next we introduce Edwin Hubble, the father of modern cosmology, and discover how he relegated the Sun and its solar system to a minor position within the Milky Way and found evidence that other galaxies existed. We also reveal why Hubble's Law – the relationship he discovered between the speed at which other galaxies are travelling and their distance from the Earth – showed that the universe is expanding and that at some point in the distant past everything started from the same point.

The most widely accepted view on how this all happened – the Big Bang – is explained in depth in this part too. We look at how the Big Bang got its name and why Einstein's equations of relativity support the idea that the universe underwent a Big Bang. We also discuss the cosmic microwave background, which is the radiation that pours down on the Earth from space that offers the best evidence that a Big Bang actually happened.

We then explain why the Big Bang on its own is not enough and how something called *inflation* – a rapid expansion of the early universe in a very small timescale – must have happened to create the universe that you inhabit today.

Finally, because the Big Bang is not the only scientific explanation of the universe, we discuss the pros and cons of the steady state, tired light, and Mixmaster theories of how everything began.

Part III: Building Your Own Universe

To build a universe from scratch requires a good toolkit and a wide range of building blocks – just what you discover in Part III.

We introduce you to *the Standard Model*, a set of fundamental particles with very odd names that make up everything you see.

We also explore the four fundamental forces of nature – gravity, electromagnetism, and the weak and strong interactions – and see how the behaviour of the fundamental particles are governed by these forces.

We then move on to *how* these basic building blocks are put together to create atoms and molecules, giving rise to the science of chemistry. We look at how the heavier chemical elements are forged in the heart of stars and the importance of this process in the birth of life itself.

From these beginnings, we finally share how stars, galaxies, and larger structures in the universe were created and how, by looking at the shape of the universe today, scientists hope to reveal what happened at the very beginning.

Part IV: Asking the Tough Questions

This part looks at some of the weirder things about the universe. Ever wondered what a black hole really is? This part explains it. We introduce you to some strange heavenly bodies. In fact, if it's strange and it's in the skies, you can find it in this part.

We also ponder the existence of alien life. Do other planets in the universe support life? The answer is a probable yes, if only because the conditions to make it happen on the Earth may have happened on any other world in the huge universe.

Lastly, we look at how the universe may end. This may seem like science fiction, but scientists have been able to work out a few scenarios of what Doomsday may really be like. Is it going to be a Big Crunch, a Big Chill, or a Big Rip? Read and decide for yourself.

Part V: The Part of Tens

This fun and quick part starts with a look at ten different explanations for the creation of the universe, ranging from what's written in the Bible to Terry Pratchett's *Discworld* universe. We also share with you ten observatories, experiments, and laboratories that can tell us a lot about the origins of the universe.

Finally, the Appendix brings together the key mathematical equations that cosmologists use to describe what happens in the universe. You can skip this section if numbers aren't your strong point. But if you like getting your hands dirty, the Appendix is the place for you. We also explain how cosmologists measure things in the universe, relying on concepts such as light years and parsecs. You have been warned.

Icons Used in This Book

If you're a cosmological geek, this icon points out all the bits that you're sure to love. If, however, algebra brings you out in a cold sweat, you're probably best to just skip these icons.

We use this icon to highlight the eureka moments – the points in history when great thinkers came up with big ideas. These times are when cosmology's and humans' understanding of the universe jumped forward in a huge leap.

This icon points out things about the universe that may take some getting used to. Many things about how the universe works and ideas on how it was created are counter-intuitive, to say the least. This icon tells you to brace yourself because you're about to encounter something very strange.

These notions are the key concepts about cosmology in the book. You may want to take a second look at these paragraphs and even commit this material to memory.

Where to Go from Here

Dip in wherever you want – that's the beauty of *The Origins of the Universe For Dummies*. Don't feel that you have to read this book from beginning to end. Whether you're interested in this history of the universe, the Big Bang theory, or the search for extraterrestrial life, we have something for everyone.

Part I

In the Beginning: Early Ideas About Our Universe

In this part . . .

Before the advent of television, humans had plenty of opportunity to look up at the stars. With so much time on their hands to observe the heavens, it's hardly surprising that early humans came up with some pretty elaborate explanations for why the night sky looks as it does. In this part, we explain some of the most popular ideas from the pre-scientific era.

Eventually, the Ancient Greeks put the study of the universe onto a scientific footing for the first time, believing they'd discovered the perfect answer for the workings of the universe, based on geometric shapes and a universe centred on the earth.

Much later, scientists eventually realised that the Greeks' idea of the perfect heavens was somehow flawed. We show how the realisation by Johannes Kepler that planets moved in orbits that aren't circles led to one of the most earth-shattering ideas of all time – gravity. With gravity, the modern science of cosmology was born.

Chapter 1

Exploring the Early Universe

In This Chapter

- Pondering the very beginning of the universe – and beyond
- Looking to science and religion for explanations
- Introducing cosmology
- Appreciating this current moment of cosmological discovery

Nothing is more human than wondering where you come from.

Just look at your average 3-year-old: They like nothing more than to embarrass their parents by asking them how they were born (and why that lady over there is wearing such a funny-looking dress).

Pondering the mysteries of the universe is also ingrained in growing children. One of the first nursery rhymes children recite is 'Twinkle, Twinkle Little Star', which contains the simple yet profound line 'How I wonder what you are'. And children's earliest attempts at art often include pictures of the Sun and the Moon – made more friendly with the addition of smiling faces, of course.

Little surprise, therefore, that by the time children are pre-teens – and sniggering over technical diagrams of human procreation – they are simultaneously starting to ask deeper questions about the skies: Why is it blue? Why do stars only shine at night? Are the Sun and the Moon the same every night or do they pile up in a discarded heap beyond the horizon? Films like *ET* and *Star Wars* only help to fuel the curiosity.

That curiosity that doesn't vanish with childhood, either. As an adult, you may find yourself pondering the multitude of stars in an especially dark night sky, or being caught off-guard by a particularly beautiful moon. If any of this sounds familiar, then *The Origins of the Universe For Dummies* is definitely for you.

Shifting Views – Scientifically Speaking

Imagine for a moment that you're a pupil at school in the first few years of the 20th century. At this time, the most famous scientist in the world is still probably Sir Isaac Newton (turn to Chapter 3 for more on Newton). Several years still need to pass before the name Albert Einstein trips off every schoolchild's tongue and the famous photo of a straggly grey-haired scientist becomes one of the most instantly recognisable images in the world (Chapter 4 tells you about Einstein).

In early 20th-century science lessons, your instructors are likely to teach you about

- **Newton's equations of motion.** The mathematical formula $F = ma$ is vital, and you're expected to know all about equal and opposite reactions (or pretend that you do anyway).
- **Electricity and magnetism.** These two forces are all the rage in the early 20th century (and are still learned in school today). You can almost certainly quote Ohm's Law (the famous $V = IR$). You also probably know all about James Clerk Maxwell's realisation that electricity and magnetism are different aspects of the same thing (see Chapter 4 for more details about Maxwell).

If these topics don't sound too difficult to grasp, wait until a few years after the end of the First World War. Suddenly, the science syllabus expands, and students are introduced to what has become the most famous equation of all time – $E = mc^2$. (Parents of these pupils, who were steeped in Newton's laws of motion at school, are suddenly on dodgy ground when it comes to helping out with the homework!)

The bottom line? Educating people about science changes all the time because science changes all the time.

For 200 years, everyone thought that Newton's views of the universe would never be bettered, and the vast majority of scientists believed that the equations he formulated described the universe in its entirety. But then along came Einstein with his crazy ideas about relativity as well as mass and energy being interchangeable. Everything changed. Yet this change wasn't an instantaneous process. Einstein published his special theory of relativity in 1905 but years passed before scientists widely accepted it.

Science works on consensus. A cherished view of how things in the universe are arranged may exist for years – even centuries – and then someone comes along and says, 'Aha. What about this?' Initially, the new idea may be dismissed, but as other scientists verify these new ideas, the consensus can change. That's why Einstein's theories are now believed to better describe the universe than Newton's.

Science is never a done deal. As you read this book, a scientist is sitting somewhere having a eureka moment, perhaps realising that Einstein's theories don't explain everything. Perhaps he or she is even beginning to realise the current theories about the origins of the universe, as described in this book, need refining. That scientist may have trouble convincing the thousands of scientists and students who do believe completely in Einstein. However, if the new theory has merit, a new consensus forms. When it does, the schoolchildren of tomorrow are going to be studying something different. And you may end up being the parent having difficulty comprehending *their* homework.

Contrasting Science and Religion

Science is one thing – religion is certainly another.

In most societies, kids are presented with religious ideas. In some cases religious and scientific ideas are in direct opposition, and in other cases they're not.

- Some people come to accept a religious viewpoint and discard any scientific notions that contradict it.
- Others come to the conclusion that science offers the most believable answers and discard any religious notions.
- Many go along in life juggling the two – accepting ideas like the rise and fall of the dinosaurs millions of years ago, while believing that God (or a god) created the world.

Growing children – not to mention inquisitive adults – often have difficulty knowing which is right: science or religion.

In many critical ways, people who preach science and people who preach religion are similar. Both ask their adherents to make spectacular leaps of faith. Christians are asked to believe that Jesus performed miracles, whereas those steeped in science are challenged with the idea that the Earth rotates around the Sun rather than the seemingly obvious opposite. Both preachers and scientists argue that proof exists of their own views of the world.

At this point, you may be thinking, 'Hey, hang on a minute. Have I picked up *Religion For Dummies* or *Philosophy For Dummies* by mistake?' A quick check of the front cover reveals not, but we can't talk about the origins of the universe without at least a nod to the fields of religion and philosophy.

Drop into any university hall of residence after midnight, follow the smells of the strong coffee, and you soon find yourself immersed in just such discussions. If one immutable law of the universe exists, it's not that everything is affected by gravity or that energy isn't created or destroyed – it's that undergraduates in higher education ponder on how it all began, just as they did in a smaller way when they were 3 years old.

In this book we try to answer some of these tough questions. (Unfortunately, the scientists' answer in many cases is that we still don't really know.) However, we didn't write these chapters just for philosophising undergraduates. We're writing to appeal to anyone who has ever wondered where everything came from.

Defining Cosmology

Cosmology is the study of the development of the universe – small word; big topic! It tries to answer questions about how the universe came to be the way it is now, and where it's heading.

The big challenge with cosmology and the related science of astronomy – the study of all the stuff out in space – is that they are unlike most other sciences. In chemistry, for example, you can add one chemical to another in test tubes that you hold in your hands. In biology, you can put a beetle under a microscope and start dissecting it there and then.

Cosmology and astronomy are different. Humanity has only ever ventured as far as the Moon – a distance of a quarter of a million miles. Although that sounds like a long way, it's nothing to the scale of the universe. The Earth's nearest star, the Sun, is 150 million kilometres away – 360 times farther than humans have ever ventured. How can humans ever hope to understand the universe if they've explored so little of it? The answer is in the science of cosmology.

The word *cosmology* comes from Greek roots – *kósmos,* meaning world or universe and *logos,* meaning word or study. Yet the word wasn't coined until long after the ancient Greeks lost their power. The term was probably first used some time in the 18th century when natural philosophers (as some scientists were then called) starting looking at Newton's work and realised that it changed humans' entire view of the universe.

Cosmology and astronomy are very closely related, but whereas astronomers study everything within the universe (stars, galaxies, and so on), cosmologists study the universe and its evolution as a whole. As a result, cosmologists need to know about astronomy as well as physics – both the traditional and the odder kinds of physics, such as quantum mechanics.

So how do you become a cosmologist? Just thinking about the origins of the universe makes everyone into an amateur cosmologist. And the purpose of this book is to help answer some of these tough questions.

If it's been a while since you thought about science, you may want to take a look at the Appendix, which outlines the special ways scientists use to describe numbers and the units of measurement that are sometimes hard to comprehend.

Seeing the beginning of the universe

Knowing how the universe began can be very helpful in understanding why the universe is the way is.

So how can cosmologists see the beginning of the universe? The short answer is that they can't – not directly at least. Sorry. If you were expecting a simple, definitive answer – like Douglas Adams' assertion in *The Hitchhiker's Guide to the Galaxy* that the answer to all the questions of the universe is 42 – you're going to be disappointed with cosmology.

But that is not to say there is no way to find out about the universe's past. That's because, as Einstein showed, 'time' isn't as straightforward as the clock on your wall suggests. This strangeness actually helps with the study of cosmology.

For example, when you look around at the night sky, you're actually looking at the universe at different stages of its development. When you observe a galaxy so far away that its light is 12 billion years old, you're essentially seeing a galaxy that was one of the earliest ever created. If cosmologists can figure out how this early galaxy was formed, they know something about how the universe was immediately before the formation. By taking similar small steps backwards, scientists can get closer and closer to the universe's starting point. See the sidebar 'When is now?' for more mind-blowing information about the nature of time.

When is now?

One of the big problems of cosmology is how to define *now.* Now is a very subjective idea.

Imagine your friend is standing at the other side of a large field holding a big red balloon. You have agreed with her that when *now* arrives, at the point when the time reaches midday, she pops the balloon.

Do you define the moment of now as

- The point when your synchronised watch shows midday?
- The point when you see your friend burst the balloon?
- The point when you hear the balloon pop?

Your friend insists that all three of these things happen at the same moment, yet your senses tell you that midday strikes, a tiny fraction of a second later you see the balloon burst, and then a couple of seconds later you hear the pop.

The same is true of the universe. When scientists observe the heavens, they're just seeing an Earth-bound version of now. If someone was to burst that balloon on the surface of a planet circling the nearby (in galactic terms) star of Proxima Centauri and you could observe the balloon's burst with a telescope, the light from Proxima Centauri would take more than four years to reach the Earth.

Furthermore, the most distant galaxies in the universe set off their light some 12 billion years ago. In the intervening time, one or more of these faraway galaxies may have exploded. Aliens living on a planet in one of these distant galaxies would certainly disagree with an Earth-based definition of now.

Dealing with the stretch limousine effect

On its own, looking farther away into space in order to see the past isn't enough to understand the origins of the universe. Even with more and more powerful telescopes that enable cosmologists to see farther and farther back in time, scientists reach a point beyond which they can never see.

At some point long ago, the universe wasn't composed of chemical elements such as hydrogen and helium gases, as it is today. Instead, the universe was made of smaller things – individual particles, such as electrons, protons, and neutrons. (Check out Chapter 9 for more on these.) In the early days these particles floated around freely, emitting and absorbing radiation.

This era, which scientists now believe ended some 380,000 years after the beginning of the universe, acts like the smoked windows on a stretch limousine. No matter how powerful the telescopes humans invent, scientists can never see through this smoked window.

This apparent barrier hasn't stopped scientists trying. In fact, cosmologists have detected a faint glow coming through this smoked window, known as the *cosmic microwave background* (see Chapter 6). By studying this glow, which over the years has cooled down to a point a few degrees above absolute zero, cosmologists hope to get a glimpse of the party going on inside.

As anyone who has ever tried to peep through the windows of a stretch limo can tell you, working out who or what's inside is almost impossible. But what scientists have discovered so far about the glow behind the universe's limo is strong evidence that the universe began in something called the *Big Bang*. In Chapter 6 we examine this amazing process, in which an infinitesimally small point expanded into the universe you see today.

Starting from scratch

If humans can't see beyond the smoked glass to the Big Bang itself, what hope do cosmologists have of understanding the origins of the universe? Luckily, scientists have devised other ways to study the origins of the universe.

You may have heard or read about places like CERN in Geneva, Switzerland or Fermilab, near Chicago, Illinois – fascinating subterranean laboratories with machines and gadgets that are extremely expensive to run. These laboratories and others are providing an alternative to trying to see beyond the smoked glass. The rationale behind these expensive endeavors is as follows: If scientists can't see beyond the glass, why not try to recreate what the first moments of the universe must have been like?

What scientists have discovered so far through CERN, Fermilab, and other projects is that the earliest universe consisted of an awful lot of particles zooming around. Everything had an incredibly high energy or temperature. By smashing together things like electrons and protons (two of the elementary building blocks of the universe) in machines called *particle accelerators* and watching what happens, scientists are replicating the earliest universe.

Based on their work, scientists believe that something very strange was going on back then. After a collision between two cars, you always end up with two cars, perhaps in a state of disrepair but recognisably automotive in nature. However, when elementary particles smash together, you don't have the same things afterwards. You may start off with two protons, but you can end up with a huge spray of other exotic particles that are created out of the energy of the collision.

Einstein's work (see Chapter 4) is essential to understand this seeming inconsistency. Specifically, Einstein's realisation that energy and mass are interchangeable helps explain how elementary particles can change into more exotic particles when smashed together. Scientists at places like CERN and Fermilab are looking at these exotic particles and figuring out what happened beyond that smoked glass.

Cosmology isn't just for scientists with access to the most powerful (and most expensive) telescopes and particle accelerators. Anyone interested enough can still make a splash. For example, three secondary students at the North Carolina School for Science and Mathematics examined public data from NASA's Chandra X-ray Observatory spacecraft and used it to discover the existence of a pulsar, a rapidly rotating star that gives off a distinctive signal. Who knows, maybe this book is your first step to discovering something new or explaining the unexplainable?!

Realising Why Now Is So Exciting

Even though scientists may have decided that 'now' is an outdated concept, the times you live in are very exciting for both amateur and professional cosmologists. Modern cosmology is in its infancy:

- A little more than a century ago, Einstein came up with his cosmos-shattering insights.
- A bit more than 80 years ago, astronomer Edwin Hubble (who we talk about in Chapter 5) showed that other galaxies besides the Milky Way existed. And if that weren't enough, Hubble also showed that the universe is expanding, which provides strong evidence that the universe began with a Big Bang.
- In the middle decades of the 20th century, *particle physics* – the physics of the tiniest particles that make up the stuff in the universe – came into being. As we discuss in Part III, understanding how these smallest pieces of matter function and interact tells scientists much about the origins of the universe.
- In the last two decades, dedicated cosmology experiments – like the COBE and WMAP space missions – started to spring up. Scientists are still trying to work out exactly what the results from these satellites mean in the grand scheme of things.

Over the next few decades, scientists are certain to find out more about the universe you live in and where it came from. Human curiosity will ensure that this happens.

If you're willing to share that curiosity, join us now for a rollercoaster ride through the cosmos.

Chapter 2

Looking Up at the Stars: Early Beliefs

In This Chapter

- Seeing the universe through ancient eyes
- Adopting a scientific approach with the Greeks
- Starting a revolution with Copernicus

In this chapter we take you on a quick side-trip through time. We know that you bought this book to read about the origin of the universe as modern science understands it, and don't worry, that's still the destination. But getting a little perspective is always useful.

For most of history, humans have contemplated the origins of the universe, coming up with conclusions that are very different to modern, science-based explanations. Considering that humans have been able to study sub-atomic particles, measure radiation, or send satellites into space only in the last century or so, earlier thinkers had no option but to base their ideas on what they saw in the sky above them. As a result, for thousands of years, people put the Earth at the centre of things. Only when Copernicus arrived on the scene in the 15th century was this Earth-centric idea seriously challenged. Copernicus's revolution paved the way for the modern era of cosmology, which we cover in Part II of this book.

This chapter ventures back to some of the earliest known human beliefs about the origins and workings of the universe and charts the gradual development of these beliefs. Read on; it's an entertaining ride.

Making a Home for the Gods: Early Notions

Gazing up at the stars on a clear night and not feeling a sense of awe is almost impossible to do. And so, unsurprisingly, civilisations throughout history have told tales to explain the way the universe was made.

In a lot of cases, people bundled their explanations with a host of weird and wonderful creation stories. In some of these tales, the Sun and Moon are gods moving through the skies, whereas other tales suggest that the night represents the underworld. The following sections explore two fascinating visions of the universe – those of the ancient Babylonians and the Egyptians, civilizations that may predate by centuries the writing of the creation stories in the Bible.

Whatever the specifics of these stories, the general view was that the Earth lay at the heart of the universe and anything that shifted in the sky moved relative to the humans below, here on the ground.

Splitting a god's carcass in two: The Babylonian creation story

One particularly gory example of a creation myth arose among the ancient Babylonian people, who lived on the plain between the Tigris and Euphrates rivers in an area occupied by modern Iraq, parts of Syria, Turkey, and Iran from roughly 2000 BC to 500 BC.

According to the Babylonian creation myth, known as *Enuma Elish,* the Earth was created after a fight between two gods – Tiamat, the monstrous embodiment of chaos, and the younger Marduk. In the battle, ferocious Tiamat opened her mouth to swallow her opponent, but Marduk seized the chance to fill her with hurricane winds. The winds filled Tiamat, leaving her vulnerable to Marduk's arrows and lances.

The story goes on to explain that Marduk – stopping only long enough to sever Tiamat's limbs, smash her skull, and slice her arteries – split Tiamat's body 'like a cockle-shell' and used the top half to construct the arc of the sky and the bottom half to make the Earth.

After his bloody victory, Marduk then found conspicuous places for all the great gods in the sky, giving them starry aspects as constellations. Furthermore, Marduk opened Tiamat's ribs to serve as gates in the east and west for the Sun to rise and set and to provide the Moon with a jewel-like lustre. The text, written around the 12th century BC, quotes Marduk's instructions to the new moon (the point every 29 or 30 days when we only see the unilluminated side of the Moon, because it lies between the Earth and the Sun and its far side is lit). Here's one translation: 'When you rise on the world, six days your horns are crescent, until half-circle on the seventh, waxing still phase follows phase, you will divide the month from full to full.' This is an accurate representation of how the phases of the Moon progress.

Gory fantasies aside, mind you, the Babylonians had a well developed mathematical system and were keen star watchers. They catalogued the movements of the stars and planets and recorded eclipses, mainly for the purposes of astrological prophesying. Their observations and predictions were surprisingly accurate given that they had little in the way of scientific instruments with which to make their recordings.

Making love among the stars: Egyptian gods in the sky

The ancient Egyptians (roughly 3100 BC to 30BC) had a complicated set of mythologies, but one of their best known myths is a creation story that starts with the primeval waters of the god Nun. (If you want to know more about the Egyptians' mythology and scientific contributions, check out *The Ancient Egyptians For Dummies* by Charlotte Booth, published by Wiley.)

The story goes that from these waters, a mound appeared upon which sat the god Atum. Atum spat to produce the gods of air, called Shu, and moisture, called Tefnut. They in turn gave birth to the god and goddess of earth (Geb) and sky (Nut).

Geb and Nut were apparently bound at first in an eternal embrace, but Shu separated them, leaving Geb frozen in eternal torment, while Nut was lifted into her place in the sky. Shu positioned himself as the air separating them. Each day, the heavenly bodies entered Nut's mouth, moved through her skies, and at dawn were reborn from her womb.

According to one version of the tale, Shu ruled that the pregnant Nut should not give birth any day of the year. The desperate Nut then pleaded to the god Thoth for help. Thoth gambled on her behalf with the moon-god Yah and won five more days to be added onto the year, which had up until then been 360 days long. Nut gave birth to one child on each of these days: Osiris, Isis, Set, Nephthys, and Horus-the-Elder. These extra days proved rather useful. Previously, with a 360-day year, priests had to declare an additional month every few years to get the year and seasons (as dictated by the movement of the Sun) back into step.

According to lore, Thoth taught the ancient Egyptians how to watch the heavens and gave them the names of 36 constellations. Careful monitoring of the sky was important stuff for the practically minded Egyptians who needed a regular calendar to help them get ready for the flooding of the Nile.

Taking a Scientific Approach: Early Views

Many ancient civilisations, like the Babylonians and Egyptians, took religious or spiritual views about their place in creation. The ancient Greeks, however, were the first to hold a distinctly different perspective on the universe.

The Greeks thought of the heavens as something that mere mortals could understand, rather than as belonging strictly to the realm of gods. In short, the Greeks took a scientific approach, laying the foundations upon which scientists continue to build much of today's understanding of the universe.

Of course, the early Greeks had their gods too. One version of the Greek creation story, for example, involved Eros, the god of love, creating order out of chaos from which night and day, and eventually the Earth itself, arose. These gods were said to reside at the top of Mount Olympus, ruling the world at their whim.

But over the centuries, Greek thinkers began to realise that the stars in the night sky offered patterns that were stable enough to use to navigate ships. This realisation gradually opened their eyes to the fact that physical laws, not the random decisions of deities, governed the stars.

The following sections contain details of some of the ancient Greeks' great physical discoveries and realisations.

Envisioning the harmony of the spheres: Pythagoras

At school you had to know Pythagoras's theorem, which lets you calculate the lengths of the sides of a triangle, but Pythagoras and his followers were fascinated by many other topics involving numbers as well. This group, which lived in a kind of religious brotherhood beginning in the sixth century BC, felt sure that the universe was deeply mathematical in nature.

Pythagoras thought the shape of the universe was based on the sphere, the most perfect geometric object according to the Greeks because it could be defined by a single parameter, its radius. He considered the Earth to be at the centre of that sphere, around which the Moon, Sun, and planets moved on their own concentric wheels. He also thought that the speeds at which those heavenly bodies moved created perfect harmony.

A follower of Pythagoras, Philolaus, later came up with one of the first recorded concepts of the universe that didn't put the Earth at its centre. Philolaus's scheme had the Earth as a perfectly spherical object, which orbited in a circle around an invisible fire along with another nine heavenly bodies (see Figure 2-1). Still, most ancient Greeks continued to believe that the Earth was at the centre of things.

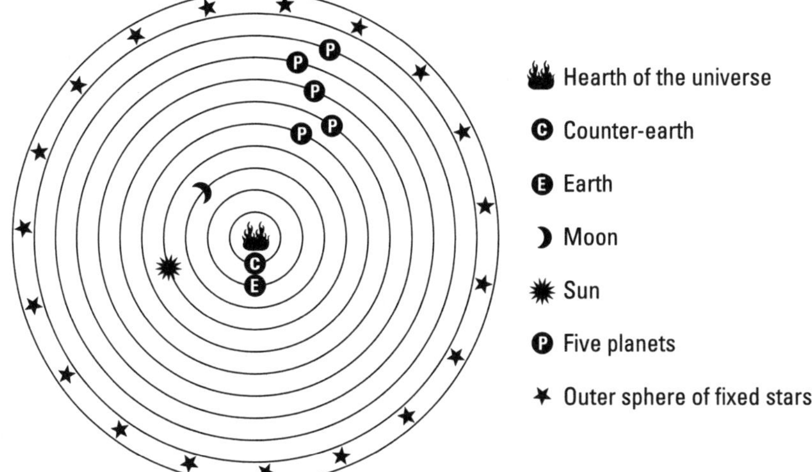

Figure 2-1: Philolaus's view of the universe.

Pushing the limits: Anaxagoras of Clazomenae

Challenging the established thinking can be dangerous, as a fellow called Anaxagoras of Clazomenae found in around 450 BC when he offered new ideas on the origins and organisation of the cosmos.

Anaxagoras challenged the religious teachings of his day, specifically by saying that the Sun was a red-hot stone (instead of a god) and that the Moon reflected the Sun's light. These two assertions, although spectacularly insightful, landed Anaxagoras in prison; but incarceration didn't stop his radical thinking. He also thought that

- The universe began with a swirling vortex, which started by separating the hot 'ether' (the fiery bits) from the cool air, giving rise to water, clouds, earth, and stones.
- The circular motion of the universe made heavy, dense material congregate in the centre to form the Earth, while the violence of the spinning caused the fiery ether to tear stones away and kindle them into stars.

Anaxagoras is also remembered for coming up with the first accurate explanation for eclipses of the Sun and Moon, based on his spot-on thoughts about the Moon reflecting the Sun's light.

Following a wandering star: Aristarchus of Samos

By the third century BC, another Greek philosopher called Aristarchus of Samos came up with a more accurate description of the way the universe works. Aristarchus based his thinking on the fact that Greek astronomers had noticed that two types of star existed:

- **Fixed stars,** whose position in the sky relative to each other stayed constant.
- **Wandering stars,** which moved around. (We get the word planet from the Greek word for wanderer, *planētēs*)

Aristarchus came up with a clever argument for explaining the two star types. First, he used geometry to make an estimate of the size of the Sun, coming to the conclusion that it must be enormously bigger than the Earth. Then he argued that no way existed for something so enormous to trail around circling the much smaller Earth.

Aristarchus proposed that the Earth must orbit the Sun. And what's more, he deduced that the Earth itself must be spinning on its axis, in order to explain the apparent movement of the stars.

Now these insights were superb stuff. Aristarchus had nailed many of the key ingredients of the so called *heliocentric,* or Sun-centred, system that Copernicus (see the later section 'Moving the Sun to the Centre: The Copernican Revolution') made famous many centuries later.

Winning the day: Aristotle

Sadly, Aristarchus's revelations were dismissed out of hand because they contradicted the views of Aristotle, who had raised a couple of pertinent questions when considering whether the Earth moved:

- If the Earth is rotating, why do objects thrown upwards fall in the same place?
- Why doesn't the Earth's rotation create really strong winds?

Aristarchus had hit upon a good description of our solar system, but his system was not widely accepted. Instead, Aristotle's views prevailed.

Aristotle's description of the way the universe works was laid out in his book *On the Heavens*. He argued that nine transparent concentric spheres encircle the Earth. The outermost sphere was the heavens, whose stars appeared in the same relative positions night after night (apart from rotating around the Earth, of course), whereas the rest contained the Moon, the Sun, and the five planets known at the time (Mercury, Venus, Mars, Jupiter, and Saturn).

Furthermore, Aristotle thought that the universe was not infinite in size because it moves in a circle; if the universe were infinite, it would be moving an infinite distance in a finite time, which is impossible. On the other hand, Aristotle said that the universe was *eternal* – in that it always has, and always will, exist.

Aristotle's views dominated Western thought until the 16th century, even though his idea of perfectly circular motions didn't really stand up to scrutiny. As the Greek empire expanded to the east, astronomical data collected by the Babylonians and Egyptians (who were both under Greek imperial rule at different times) became available to the Greeks. These records clearly indicated that the planets didn't move in circles around the Earth at all.

Refining Aristotle's system: Ptolemy

For centuries – more than a millennium, in fact – no one proposed any real challenges to Aristotle's model for the universe.

But in the second century AD, a Greek astronomer called Claudius Ptolemy, who was born in Egypt and had Roman citizenship, added some refinements that made the Aristotelian scheme do a much better job of matching the movements of the planets across the skies.

Ptolemy's first clever move was to move the Earth just slightly away from the centre of the cosmic spheres. This shift helped explain why the planets seemed sometimes to move closer to the Earth or farther away.

Another of Ptolemy's achievements was to explain the odd movements of some planets, such as Mars, which seem to backtrack on themselves as they move across the sky – something referred to as *retrograde motion*. This backtracking in fact happens because the Earth orbits around the Sun more quickly than planets farther away, but Ptolemy had to figure out an explanation that had the Earth standing still.

His idea was that each planet moved around in a series of small circles, each of which in turn spun on larger spheres. At least one writer has likened this arrangement to a kind of Ferris wheel where capsules spin around on the big wheel. These systems of smaller circles were called *epicycles*. To explain the motion of the planets, Ptolemy added a complex set of supplementary orbits to the movement of some planets.

Although Ptolemy's system was actually pretty good at explaining and predicting the movement of heavenly bodies, it wasn't accurate. Of course, Ptolemy's aim wasn't really to describe the physical reality of the universe, only to find a way to chart the movements of the planets, Sun, and Moon. For more on Ptolemy's work, see the sidebar on page 25.

Moving the Sun to the Centre: The Copernican Revolution

Polish church canon Nicolaus Copernicus kicked off a revolution in cosmology when he presented a model of the world in which the Sun was central, not the Earth (see Figure 2-2). Although Aristarchus had done something similar many centuries before (refer to the earlier section 'Following a wandering star: Aristarchus of Samos' for more information), Copernicus's model was the one that eventually led to a change in the way humans view the universe.

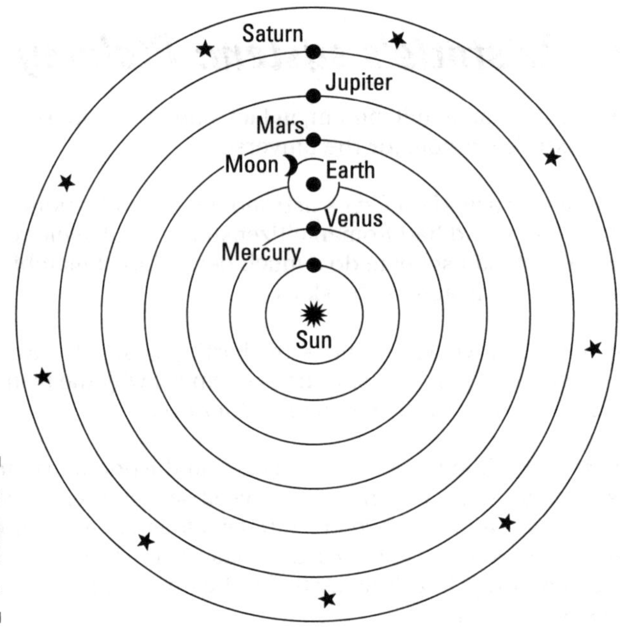

Figure 2-2: Copernicus's Sun-centred universe.

Copernicus first set out his plan on a short handwritten note known as the 'Little Commentary', or 'Commentariolus', which he circulated among friends and colleagues between 1508 and 1514. He wrote that if the Sun is assumed to be at rest and the planets are orbiting it, you can calculate the planets' relative

Writing the great treatise

Ptolemy's great astronomical breakthroughs are collected in a work called the *Almagest,* also known as the *Syntaxis.* The word *almagest* is a Latin form of an Arabic name *al-kitabu-l-mijisti,* which means 'The Great Book'.

As its name suggests the *Almagest* brought together all mathematical astronomy of the day into a single work, including earlier Greek astronomy as well as Ptolemy's new work. The work is an amazing collection of theoretical and observational information, from geometry to observations about the Sun, the Moon, and the five planets that are visible to the naked eye.

The work was accepted as definitive soon after Ptolemy finished it, some time around the year 150. In fact, for many centuries, his explanation of the universe was the runaway favourite among star-watchers.

distances from the Sun based on the length of time it takes them to cross the sky. The circular orbit of the Earth around the Sun also explained the fact that the planets appear to change in size and helped make sense of the retrograde motion of some planets.

The 'Commentariolus' contained seven assumptions or axioms that underpinned Copernicus's ideas:

- **The heavenly bodies don't all move round the same centre.** A key piece of evidence would be if anyone observed a Moon around another planet, as Galileo later did around Jupiter.
- **The Earth isn't the centre of the universe, only of the Moon's orbit and of terrestrial gravity.** Heretic stuff, of course, but it would dramatically simplify the maths of orbits.
- **The Sun is the centre of the planetary system and therefore of the universe.** This laid the foundations for the scientific understanding of planetary motion (that is, gravity).
- **Compared to the distance between the Earth and the fixed stars, the distance from the Earth to the Sun is negligibly small.** Finally, someone had realised that the universe was immense.
- **The apparent daily revolution of fixed constellations (as opposed to those pesky wandering planets) is due to the Earth's rotation on its own axis, the imaginary line joining the north and south poles.** This was a revolutionary, counter-intuitive idea, but very, very clever.
- **The apparent annual motion of the Sun is due to the fact that the Earth, like other planets, revolves around the Sun.**
- **The apparent 'stations' and retrograde motion of the planets are due to the Earth and planets revolving around the Sun.**

The final version of Copernicus's theories didn't appear for another 30 years or so, when, in the year of his death, 1543, he published his treatise *De Revolutionibus Orbium Coelestium Libri VI* (*Six Books Concerning the Revolution of Celestial Spheres*).

This book, set out with mathematical proofs, provided a system that was as good as Ptolemy's at making astronomical predictions, although Copernicus's system still relied on *epicycles* – refer to the preceding section 'Refining Aristotle's system: Ptolemy' – and envisioned circular orbits for the planets. It also prepared the ground for Galileo's observations, with the newly invented telescope. Those observations would prove that we were not at the centre of everything after all.

In fact, the planets in the solar system orbit around the Sun in the shape of an *ellipse* – a kind of squashed circle. This revelation, based on an understanding of gravity (thanks to Sir Isaac Newton), required the work of another genius, Johannes Kepler. Chapter 3 discusses Newton, Kepler, and gravity in detail.

Nicolaus Copernicus, cautious author

Nicolaus Copernicus was born in 1473 in the town of Toruń what is now northern Poland. His father, a wealthy magistrate, died when Nicolas was just 10 years old, and so he and his siblings came under the care of their uncle Lucas Watzenrode, a canon of Frauenburg Cathedral and later bishop of Warmia.

Under the care of Lucas, Nicolaus attended university at Cracow, Bologna, and Padua, finally becoming a doctor of canon law at the age of 30, in 1503. In the midst of all these studies, his uncle arranged for him to be appointed a canon of Frauenburg Cathedral, which entitled him to a good income, servants, and horses.

Throughout his adult life Copernicus served a variety of roles, as personal physician and diplomat for his uncle, and as a regional administrator. As a church canon, he had duties that included collecting rents from church-owned lands, securing military defences, and caring for the medical needs of the other canons. He conducted his astronomical work in his spare time.

By all accounts he was a secretive and solitary man who preferred the solitude of his home and observatory, in a tower of a fortified wall in Frauenburg. He lived here for the last 30 years of his life, taking measurements of the stars and studying the works of ancient Greek astronomers.

After he circulated the brief outline of his hypothesis on the motions of the planets around 1510, called the 'Commentariolus' (Little Commentary), Copernicus became increasingly well known among learned people throughout Europe, many of whom urged him to publish his full work. He resisted for years, however. Only with the assistance and encouragement of his student, Georg Joachim Rheticus, did he arrange to have his full work printed. According to legend, the completed book arrived from the printers on the day its author died.

Chapter 3

The Apple Drops: Newton, Gravity, and the Rotation of the Planets

In This Chapter

- Observing the heavens with and without telescopes
- Pondering the possibility of an imperfect universe
- Considering the contributions of Brahe, Kepler, and Galileo
- Seeing farther with Newton

Since the dawn of time, humans have looked up to the heavens in awe at their perfection. The ancient Greeks were fascinated by the perfection of the circle (refer to Chapter 2 for more on their theories) and, quite logically, believed that the planets moved in perfect circles as well.

Unfortunately, planetary orbits aren't quite so perfect when examined in close-up. With the advent of accurate scientific instruments in the late 1500s and the invention of the telescope in the early 1600s, one fastidious early astronomer and his equally famous assistant realised that the paths of the planets weren't circles at all, but another geometric shape altogether – the ellipse. We tell you more about these two later in the chapter.

The bombshell of this discovery marked the beginnings of physical, rather than metaphysical, explanations of how the universe worked. Over the next hundred years, great scientists such as Johannes Kepler, Galileo Galilei, and Isaac Newton discovered the laws that made humans challenge the very basis of creation. This chapter explores the contributions of these remarkable men.

Tycho Brahe, a Rising Star

Denmark's Tycho Brahe was the last great astronomer of the pre-telescope era. He died in 1601, seven years before Hans Lippershey filed for his telescope patent, but that didn't stop Brahe from observing the heavens in incredible detail.

Born on 14 December 1546 at the castle of Knutstorp in the region of Scania (then Denmark but now part of Sweden), Brahe's interest in astronomy was kick-started by a solar eclipse in 1560. Tycho was something of a perpetual student, studying at universities in Copenhagen, Leipzig, Wittenberg, Rostock, and Basel in subjects as diverse as philosophy, law, humanities, science, and rhetoric.

All this time, Brahe pursued his interest in astronomy. On 11 November 1572, an event occurred that changed the course of his life forever. As Brahe himself wrote:

> *I was contemplating the stars in a clear sky. I noticed that a new and unusual star, surpassing the other stars in brilliancy, was shining almost directly above my head; and since I had, from boyhood, known all the stars of the heavens perfectly, it was quite evident to me that there had never been any star in that place of the sky, even the smallest, to say nothing of a star so conspicuous and bright as this. I was so astonished of this sight that I was not ashamed to doubt the trustworthiness of my own eyes. But when I observed that others, on having the place pointed out to them, could see that there was really a star there, I had no further doubts. A miracle indeed, one that has never been previously seen before our time, in any age since the beginning of the world.*

What Brahe had observed was a *supernova,* the catastrophic explosion when a massive star reaches the end of its life. Supernovae had in fact been witnessed before – one in China in the year 185 and another in 1054, which left behind the supernova relic known today as the Crab Nebula. (Turn to Chapters 12 and 18 for more on supernovae.)

Brahe's account of the supernova in his publication *De Stella Nova* in 1573 secured his astronomical reputation and coined a new word – *nova* – for a suddenly brightening star.

A nova is a different kettle of fish from a supernova. In modern astronomy, a *nova* is a star that brightens sharply and unexpectedly but whose cause is very different from that of a supernova. Novae occur in *double star systems*, in which a small *white dwarf star* (a compact remnant that remains when a Sun-like star reaches the end of its life) orbits a larger companion star. Material, principally hydrogen, from the larger companion is drawn towards the smaller star into a large disc, which gets hotter and hotter until a runaway reaction occurs. This reaction causes the disc material to be ejected at high speed and with a brilliant burst of light.

Working in the greatest observatory ever built

Shortly after the publication of Brahe's experiences, he became a favourite of Denmark's King Frederick II who lavished huge sums on building Brahe an observatory on the island of Hven. The Tycho Brahe museum stands there to this day.

Conceptualising the Tychonic universe

Brahe's observations convinced him that he needed to formulate a new system of the universe. He realised that Ptolemy's system (in which everything orbited the Earth) was flawed, but he was also unhappy with the Copernican system, in which everything orbited the Sun (refer to Chapter 2 for more on these early astronomers).

As a result, Brahe developed his own hybrid *Tychonic* system in which the five planets known at the time – Mercury, Venus, Mars, Jupiter, and Saturn – orbited the Sun, but the Sun and Moon orbited the Earth.

The Tychonic system never achieved wide acceptance. The system failed to improve much on the orbits already predicted by the Ptolemaic and Copernican systems, which proved to be a crucial disadvantage in trying to get Brahe's theory accepted.

Brahe's other observations were equally universe-shaking. Brahe was particularly interested in Mars's orbit and measured the position of Mars in the sky when it was in opposition. A planet is in *opposition* when the Earth sits on a straight line between the planet and the Sun. The figure shows the position of the planet Mars when it is in opposition (according to our current concept of the solar system). By definition, it's a term that you can only apply to a planet farther out from the Sun than Earth is.

Despite his best efforts, Brahe was unable to make Mars's positions fit onto a circular orbit. (However, based on Brahe's observations, his assistant Johannes Kepler developed the idea of elliptical orbits.)

The death of Brahe's benefactor King Frederick II in 1588 proved harsh. His successor Christian IV had no taste for Brahe's expensive dabblings and Brahe was forced to leave the island of Hven in 1597. Nevertheless, he didn't join the ranks of the unemployed. Brahe took his instruments and records to Prague, where he became imperial mathematician to the Holy Roman Emperor Rudolph von Hapsburg.

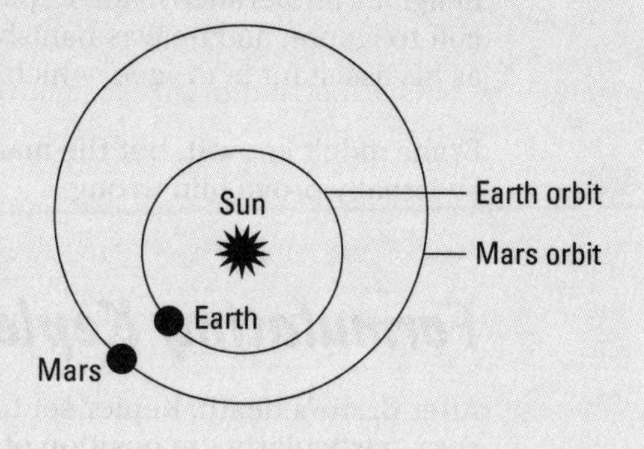

The observatory was named Uraniborg, or castle of Urania, after Greek mythology's muse of astronomy. It featured 28 instruments of Brahe's own devising, including some which, by some accounts, were more than a hundred times more accurate than anything previously constructed. Brahe used these devices to measure the positions of more than a thousand stars. Specifically, he observed the path of the great comet of 1577 and proved that its orbit was beyond the Earth's atmosphere, farther than the Moon, and crossed the orbits of the other planets. This discovery was the first evidence that the planets didn't rotate on great glass spheres as some people had previously thought. It also showed that comets weren't phenomena in the Earth's atmosphere (as thought by Aristotle and others) but were celestial bodies that were part of our wider solar system.

Assisting – and Surpassing – Brahe: Johannes Kepler

Johannes Kepler, the sickly son of a poor soldier, was born in the German city of Weil der Stadt on 27 December 1571. In 1589, he received a scholarship to the University of Tübingen. After graduation he moved to Graz to become a teacher of mathematics. He also earned extra money by making astrological predictions for the nobility.

In Graz, Kepler wrote a book entitled *Mysterium Cosmographicum* in support of the Copernican view of the universe. Tycho Brahe was one of many who read the book, and the two men began corresponding.

Meanwhile, things were changing in the staunchly Catholic city of Graz. Religious intolerance made Kepler's Lutheran upbringing increasingly difficult to ignore, and he was banished. However, Brahe offered Kepler a post as his assistant in Prague, which Kepler accepted.

Brahe didn't know it, but the man who joined him as an assistant was to eventually prove him wrong.

Formulating Kepler's laws

After Brahe's death, Kepler set to work on Brahe's meticulously gathered data, particularly the position of Mars.

Kepler picked out the best 12 points from the data and tried to find a circle that went through them all. Although this sort of task is something that a computer can handle with ease today, Kepler had to rely strictly on trial and error – and many trials and errors did he make.

Brahe's odd life and death

Although the fact has little to do with his work on the origins of the universe, Tycho Brahe is remembered for not having a nose.

An account of Brahe's life by Pierre Gassendi in 1654 reveals that Brahe lost the front part of his nose in a duel with a Danish nobleman. The duel was not over the hand of a lady – as is always the case in films – but over a mathematical dispute.

Brahe had a replacement nose made from an alloy of gold and silver, which he attached to his face with glue. We hope he'd lost his sense of smell in the process!

Life with a metal nose must have been strange, but some accounts of his death are equally bizarre. One such account, in John Allyne Gade's *The Life and Times of Tycho Brahe,* says that Brahe was invited to dinner in October 1601 by a nobleman, Baron Peter Vok von Rosenberg. In a bit of a rush, he forgot to go to the loo before sitting down for dinner. Gade's account continues:

> *Owing to the strict etiquette of the day, he did not like to leave the table, and in staying burst something of importance inside his lower regions. When he was able to totter up, it was too late, the harm was done, and his bladder wronged beyond repair.*

Brahe died a few days later. His last words, according to his assistant's records, were *'Ne frustra vixisse videar'* ('May I not seem to have lived in vain').

Although this account of Brahe's death is rather morbidly interesting, in recent years, other causes for his death have been put forward. Some people speculate that Brahe actually died of mercury poisoning. In 1991, the director of the Czech National Museum gave the Danish ambassador a gift – a box containing a fragment of shroud and a portion of Brahe's beard, which had been removed from Brahe's grave on the third centenary of his death in 1901. Scientists at the Institute of Forensic Medicine at Copenhagen University analysed the beard sample and found it contained high levels of mercury. Another analysis five years later backed up these findings and suggests that Brahe ingested mercury the day before his death. This may sound like he was murdered, but Brahe was known to dabble in making his own medicines. Some suspect that he mixed up a medicinal cocktail containing mercury to deal with an infection in his urinary tract. His concoction may well have killed him.

After extensive calculations with Brahe's data, Kepler arrived at a series of laws, which continue to affect modern understanding of the universe.

Kepler's first law

After several years of struggle trying to fit a circle to Brahe's data, Kepler tried something else. What if the orbit were not a circle at all? The great comet seen by Brahe in 1577 showed that heavenly bodies need not follow circular paths. Eventually, Kepler concluded that the orbit of Mars wasn't a circle – but an *ellipse* with the Sun at one of its foci, as shown in Figure 3-1.

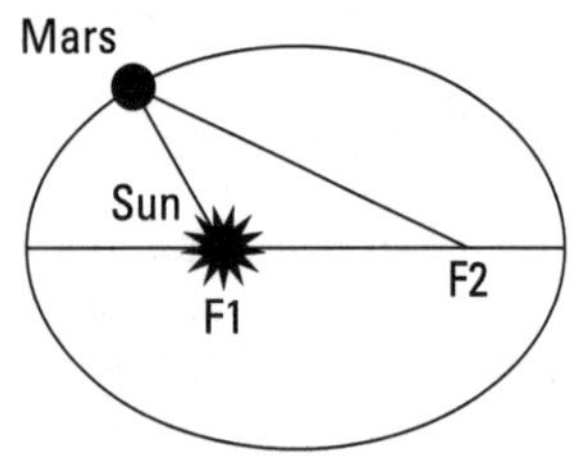

Figure 3-1: Kepler's first law.

The reason no one had spotted that planets follow elliptical paths is that orbits are only just elliptical. If an ellipse's two foci are very close together, the resulting ellipse can be very nearly circular – whereas if the foci are far apart, the ellipse can look like a squashed cigar.

Astronomers call the squashedness of the ellipse its *eccentricity*. Eccentricity in astronomical terms means the ratio of the distance from the centre of the orbit to the focus and half the diameter at its widest point and is usually expressed as a decimal fraction. A circle has an eccentricity of zero while Pluto's orbit has an eccentricity of 0.25. Comets, which also travel in elliptical orbits and return to near the Sun on a regular basis, have orbits with high eccentricities. For this reason comets only return after long periods of time. Halley's comet, for example, takes 76 years to travel round its orbit, which has an eccentricity of 0.97.

Kepler looked at the orbits of the other planets and realised that they fit the same pattern – they travel in elliptical paths with the Sun at one focus. This observation is now known as *Kepler's first law of planetary motion,* details of which Kepler published in 1609 in a book entitled *Astronomia Nova*, or *New Astronomy*.

You may be asking, if the Sun is at one of the foci of this elliptical path, what's at the other focus? The answer is: Nothing apart from the odd bit of space dust. Although imaging something solid like another star as the second focus seems beautifully symmetric, a substantial second focus would create gravitational interactions that would disrupt the planet from its elliptical path.

Kepler's second law

In addition to elliptical orbits, Kepler noticed something else: The planets didn't appear to move at the same speed all the time. In fact, he noticed that when they are close to the Sun, planets move faster than when they are at the opposite side of their orbits.

Kepler theorised that another law governed how fast planets travel. Today this law is usually stated as follows: A line joining a planet to the Sun sweeps out equal areas in equal times. Figure 3-2 illustrates Kepler's second law in greater detail.

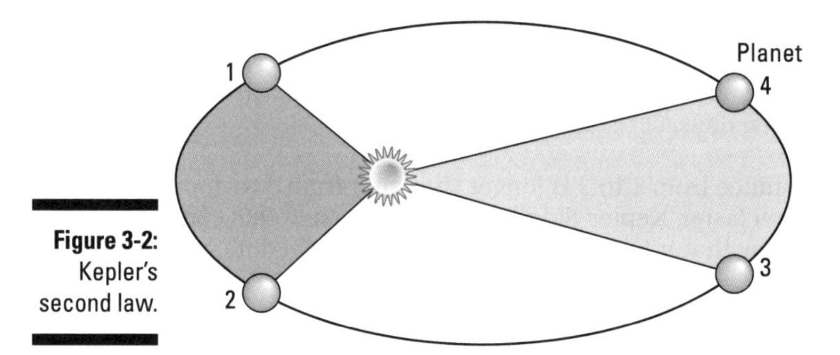

Figure 3-2: Kepler's second law.

Think of Kepler's second law this way: Imagine that the travel time for a planet to get from point 1 to 2 is ten days. During this time, a line joining the planet to the Sun sweeps out the dark grey area, as shown in Figure 3-2. You can calculate this line if you know a couple of pieces of information about the ellipse, such as its eccentricity and the distance between the foci or the width and height of the ellipse at its fattest points.

Explaining ellipses

An ellipse is a rounded shape that's a conic section. A *conic section* is not the area in a shop where you can pick up funny reading material, but rather the different types of curves you get if you slice through a cone.

Cutting through a cone at different angles produces different shapes, ranging from the well known – circles – to lesser known curves such as the parabola and hyperbola, as you can see in the figure.

An ellipse is the mathematical term for a particular type of oval. You can draw one yourself on a piece of paper by putting two drawing pins through it. Each point where you place a pin is called a focus of the ellipse (foci, in plural). Tie a piece of string into a circular loop and then pass it round the two pins. Put a pencil inside the loop and stretch the string until it's taut and then start drawing, keeping the string taut at all times. Eventually, your pencil returns to its starting point and you have a closed curving shape called an ellipse.

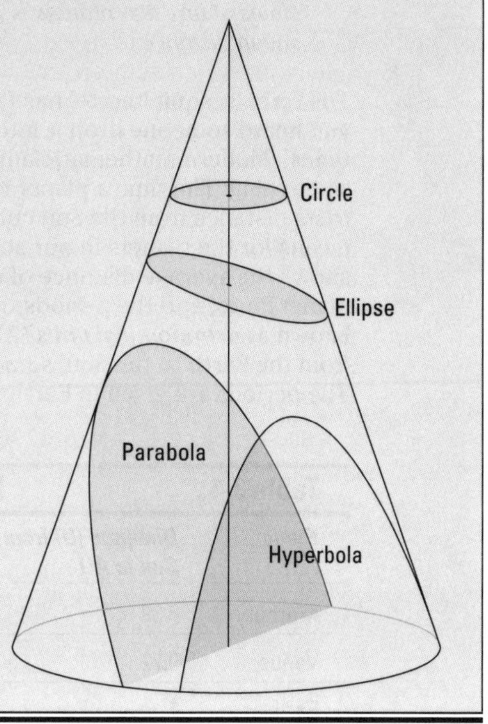

Now look at the light grey triangular wedge in Figure 3-2, which is much longer and thinner than the dark grey one but has the same area. Kepler's second law stated that the time required to get from 3 to 4 in this case was the same as from 1 to 2 – ten days.

Because the distance from 1 to 2 is longer than that from 3 to 4, obviously the planet must travel faster. Kepler didn't know what caused this change in speed, but he knew that it happened – Brahe's data proved it.

Kepler's third law

Kepler continued studying Brahe's data for another nine years before realising something else fundamental: A fixed relationship exists between how long a planet takes to complete one orbit (known as its *period* – Earth has a period of one Earth year or 365.25 Earth days; the 0.25 bit is the reason we need leap days every four years) and the average distance to the planet. This fact is not immediately obvious because the distance to the planet changes throughout the course of its elliptical orbit.

Kepler explained the relationship – which has, perhaps inevitably, become known as Kepler's third law – in his 1619 book *Harmonice Mundi,* or *The Harmony of the Worlds:*

> *It is absolutely certain and exact that the proportion between the periodic times of any two planets is precisely the sesquialterate proportion of their mean distances.*

The term 'sesquialterate' has fallen a little out of use – when was the last time you heard someone drop it into conversation? – but it means 'one and a half times'. Modern mathematicians choose a different way of expressing the same thing: The time a planet takes to orbit the Sun squared, is proportional to its distance from the Sun cubed. Sounds impressive, doesn't it? What this means for the planets in our solar system is shown in Table 3-1. The table shows the average distance of each planet from the Sun, including the dwarf planet Pluto, and the periods of their orbits. The distances are given in units known as *astronomical units* (AU). One AU is defined as the average distance from the Earth to the Sun, some 93 million miles or 150 million kilometres. The periods are given in Earth years.

Table 3-1 Kepler's Third Law in Action

Planet	*Distance (D) from Sun in AU*	D^3	*Period (T) in Earth Years*	T^2
Mercury	0.39	0.06	0.24	0.058
Venus	0.72	0.37	0.62	0.384
Earth	1	1	1	1

Planet	*Distance (D) from Sun in AU*	D^3	*Period (T) in Earth Years*	T^2
Mars	1.5	3.4	1.9	3.6
Jupiter	5.2	140.6	11.9	141.6
Saturn	9.5	857.4	29.5	870.2
Uranus	19.2	7077.9	84	7056
Neptune	30.1	27270.9	164.8	27159
Pluto	39.5	61629.9	248.6	61802

As you can see, the numbers in the third and fifth columns are virtually identical – well, close enough for scientists to believe that something significant is happening here. If we'd measured the distance from the Sun in kilometres, for example, the numbers would be different but the relationship between D^3 and T^2 is the same for every planet.

Appreciating Kepler's legacy

Despite his heavy workload analysing Brahe's data, Kepler still found time for other studies, and he has an impressive list of firsts to his name. Specifically, Kepler was the first person to

- Explain the concept of magnification of an image, as well as how pinhole cameras and telescopes worked
- Develop eyeglasses for the treatment of near and far sightedness
- Explain how humans use both eyes to perceive image depth
- Suggest the use of *parallax* (see Chapter 5) to measure the distance to the stars

Kepler also developed a theory that the Moon causes the Earth's tides. Although Galileo (who, at three years younger, was a contemporary of Kepler's) ridiculed this suggestion, Kepler was eventually proved correct.

Kepler's three laws of planetary motion are perhaps his greatest contribution. They put human understanding of the universe on a sound scientific footing for the first time. No longer did the planets whizz about the heavens on the whims of the gods. Instead, the planets followed mathematically defined paths that simple laws were able to calculate.

Despite his advances, Kepler made no guesses as to what causes the planets to act in this way. That explanation took another 50 years.

The Universe Reveals Itself: Galileo

Galileo Galilei – his name made famously musical in Queen's classic song 'Bohemian Rhapsody' – was born close to Pisa on 15 February 1564. When he entered university in Pisa, Galileo originally studied medicine but changed his course to philosophy and mathematics.

Modern science's understanding of gravity, planetary phrases, and optics is indebted to Galileo, as the following sections discuss.

Falling for gravity

In 1592, Galileo became professor of mathematics at the University of Padua and began carrying out his experiments on falling objects. The story of him dropping balls off the Leaning Tower of Pisa is almost certainly apocryphal, but he did discover that two balls, identical in size but made of different materials – iron and wood, for example – released from the same height at the same time reach the ground at the same instant, even though they have different weights.

He also noticed that objects under *free fall* (that is, objects that don't have anything to slow them down, such as a parachute) accelerate. After exhaustive experiments in which he observed objects rolling down slopes, Galileo came up with several equations of motion, including the following for the distance travelled, x, under free fall where a is the acceleration and t is the time that the object has been in free fall:

$$x = \frac{1}{2}at^2$$

Notice that the equation doesn't include a term for the weight of the ball. So when the acceleration of two objects is the same, as is the case with gravity, the two objects take the same time to travel the same distance.

Seeing through Galileo's eyes

Many believe that Galileo invented the telescope. As it happens, the telescope was patented by Dutch instrument maker Hans Lippershey in 1608. (Actually the word *telescope* wasn't coined until 1611; Lippershey called his invention a *kijker,* or 'looker' in Dutch.)

When Galileo heard about Lippershey's work, he decided to construct his own telescope. Galileo's superior version was able to magnify objects up to 30 times (probably quite a lot more than Lippershey's could) and was an essential part of his observations.

Galileo's writings were the first to suggest that not everything in the heavens was perfect, as religion had led people to believe. Specifically, Galileo's 1610 book *Sidereus Nuncius* (*The Starry Messenger*) includes belief-shaking observations about the following:

- **The Moon.** The Moon was an obvious first choice for observation. Until Galileo turned his instrument on the Moon, people had assumed that the only details on the Moon's surface were large dark spots (believed incorrectly to be seas). However, Galileo observed that the Moon's surface

  > *is not smooth, uniform, and precisely spherical as a great number of philosophers believe it (and the other heavenly bodies) to be, but is uneven, rough, and full of cavities and prominences, being not unlike the face of the Earth, relieved by chains of mountains and deep valleys.*

- **The galaxy.** Galileo looked at the nebulous clouds of the Milky Way and observed that it was

  > *nothing but a congeries* [aggregation] *of innumerable stars grouped together in clusters. Upon whatever part of it the telescope is directed, a vast crowd of stars is immediately presented to view. Many of them are rather large and quite bright, while the number of smaller ones is quite beyond calculation.*

- **Jupiter's moons.** Galileo chanced to look at the planet Jupiter and saw a group of small bright stars close to it. When he had another opportunity to view Jupiter, he was astonished to see that the stars had moved in relation to Jupiter. Repeated observations revealed four such stars, which altered their positions from one side of Jupiter to the other, sometimes disappearing altogether from view.

  Galileo took this observation to mean that the 'stars' were in fact moons orbiting Jupiter; their regular disappearances were caused by their passage behind or in front of Jupiter. Galileo used this observation as support for the Copernican, or Sun-centred, universe. The moons, called Io, Europa, Ganymede, and Callisto, are now known as the Galilean moons of Jupiter in his honour.

Going through a phase

You've heard of a full Moon but have you ever heard of a full Venus? No, we thought not.

However, in the same way that the Moon passes through phases from new to full, by way of crescent, half, and gibbous, so does Venus. You can't see the phases of Venus clearly with your naked eye – although the Mesopotamians are thought to have noticed the planet's changes. But with a good magnifying telescope or powerful pair of binoculars, you can pick out the changes.

The phases are caused by Venus orbiting the Sun – and Earth's position outside that orbit.

- When Venus is in *inferior conjunction* (Venus is sitting on a straight line joining the Earth and the Sun), you see a new Venus, because the face of Venus that is turned away from the Earth is the one that is lit by the rays of the Sun.
- When Venus is in *superior conjunction* (on the far side of the Sun on the extended straight line joining the Earth and the Sun), the lit side is pointed towards the Earth. You may expect to see a full Venus, but in fact, because the Sun is in the way, you can't.

  You can, however, see a 'new' Venus when the planet passes either above or below the Sun, as seen from Earth. You can also see it a couple of times each century, when the dark face of Venus passes across the face of the Sun (as it will next in June 2012). (Never observe the Sun directly, of course: Use an appropriate indirect viewing method – for example, view the image from a telescope projected onto a plain white card.)

If you look at Venus over the course of its year, you see a change of phases as shown in Figure 3-3. Note also that the image of Venus grows larger and smaller as the planet moves towards and away from the Earth in its orbit around the Sun.

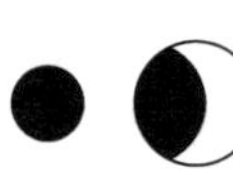

Figure 3-3: The phases of Venus.

The experience of phases may seem obvious now that scientists know that both Venus and the Earth orbit the Sun in near circular orbits, but at the time of Galileo, when the debate between the Ptolemaic, Copernican, and Tychonic systems was in flow, this fact wasn't obvious at all. In fact, Galileo's observation of these phases in 1610 showed for the first time that the Ptolemaic system was at fault.

Noting the Sun's imperfections

Galileo studied the Moon in great detail, but what of that other great heavenly body – the Sun? What drove Galileo to focus his attentions on the Sun is unclear, especially when the use of the telescope opened up such a rich collection of other sights. Yet Galileo studied the Sun extensively, probably at the expense of his own eyesight.

Galileo wrote of his observations of the Sun in *Letters on the Sunspots,* written to Marc Welser in 1612. Welser was a member of the Senate in Augsburg, Germany, and a highly regarded scholar who corresponded with many of the great thinkers of the 1600s. In the correspondence, Galileo talked of dark spots he had observed on the face of the Sun with his telescope and how they revealed that the Sun was rotating on its axis, taking around a month to complete a rotation.

In fact, Galileo was not the first to observe sunspots – that honour goes to Englishman Thomas Harriott – but Galileo was the first to suggest that sunspots were features of the solar surface rather than planets closer to the Sun than Mercury.

Galileo versus the Church

Galileo's telescopic observations made him powerful enemies. The more new discoveries he made that backed up his favoured Copernican theory of the universe, the more the Catholic Church began to view his publications as heresy. The growing popularity of Copernican theory eventually forced the Catholic Church to act.

In 1613, Father Niccolo Lorini, a Dominican friar and a professor of church history, said that the Copernican system challenged the writings of the Bible, particularly Isaiah 40:22 – 'It is He that sitteth above the circle of the Earth . . . that stretcheth out the heavens as a curtain.'

Galileo responded by writing a letter to Benedetto Castelli, a maths professor at the University of Pisa, setting out his views on the relationship between his observations and the scriptures. He argued that the Bible should not always be taken literally and wrote:

> *But that the same God has endowed us with senses, reason, and understanding, does not permit us to use them, and desires to acquaint us in any other way with such knowledge as we are in a position to acquire for ourselves by means of those faculties, that it seems to me I am not bound to believe, especially concerning those sciences about which the Holy Scriptures contain only small fragments and varying conclusions; and this is precisely the case with astronomy, of which there is so little that the planets are not even all enumerated . . .*

Three years later, Galileo was summoned before Cardinal Roberto Bellarmino by order of Pope Paul V and warned to 'abandon that opinion (that the Sun is immobile at the centre of the world and that the Earth moves); and, if he refuses to obey, that he be ordered to stop teaching, defending, and even discussing this doctrine'.

Shortly afterwards, the Inquisition denounced the Copernican system and banned Galileo's book, subject to it being 'corrected' to say that it was merely a hypothesis and not a true reflection of the state of the universe.

Galileo's contrarian views didn't stop him from maintaining friendships with leading figures in the Church. In 1623, one such friend, Cardinal Maffeo Barberini, was named the new Pope (Urban VIII) following the death of Pope Gregory XV. Galileo had lengthy discussions with the new Pope and his cardinals and was permitted to write about the Copernican theory as long he treated it as a hypothesis.

With some sense of mischief, Galileo published *Dialogue Concerning the Two Chief World Systems* in 1632. In it, two characters discuss their views of the universe with a third: Salviati (who represents Galileo's beliefs), a philosopher known as Simplicio (who follows the Earth-centred view of the universe), and an open-minded and educated friend Sagredo. In the work, Salviati explains how the Copernican model gives rise to the Earth's seasons, the variable length of the day, and other checkable observations. In the end, Sagredo is convinced of Salviati's arguments and states:

> *There is a great difference between the simplicity and ease of effecting results by the means given in this new arrangement and the multiplicity, confusion, and difficulty found in the ancient and generally accepted one. For if the universe were ordered according to such a multiplicity, one would have to remove from philosophy many axioms commonly adopted by all philosophers. Thus it is said that Nature does not multiply things unnecessarily; that she makes use of the easiest and simplest means for producing her effects . . .*

Unfortunately for Galileo, many readers of the work believed Simplicio represented the Pope. The work was banned, and Galileo hauled before the Inquisition. After two weeks of heavy interrogation, Galileo agreed to plead guilty. He was sentenced to house arrest for an indefinite term and lived out his later years at his villa in Florence. Five years after being sentenced, he became totally blind.

Amazingly, not until 1992 did the Catholic Church formally accept that Galileo had been right all along.

Watching Apples Fall: Isaac Newton

Isaac Newton was born at Woolsthorpe near Grantham in Lincolnshire in 1642, just over a decade after the death of Johannes Kepler and a year after Galileo died. Newton entered Trinity College at Cambridge University in 1661, studying mathematics. He seems to have had little exposure to mathematics and science before entering university but once there devoured all the great classic books on these topics.

The arrival of the plague in Cambridge in 1665 forced Newton back to rural Lincolnshire for three years, but this time turned out to be his most fruitful and creative. At home Newton came up with earth-shattering ideas about motion and gravity that form the basis of *Principia,* one of the greatest scientific works of all time.

The following sections cover Newton's greatest contributions to modern understanding of the origin of the universe.

Tying it all together

In 1684, Newton was visited by the Astronomer Royal Edmond Halley (of comet fame). Halley, along with the scientist Robert Hooke and architect Christopher Wren (who enjoyed a bit of science when he wasn't designing St Paul's Cathedral), had been wondering what the path of a planet would be if some force were acting towards the direction of the Sun that followed an inverse square law (see the sidebar 'Inverse square laws').

Newton responded by sending Halley a manuscript in November 1684, entitled *De motu corporum in gyrum (On the motion of bodies in an orbit),* referred to more often as simply *De motu.* In *De motu,* Newton not only answered Halley's question but also proved much more – he derived Johannes Kepler's three laws of planetary motion mathematically.

Inverse square laws

Try this little thought experiment: Imagine you have a light bulb that doesn't need to be screwed into a light socket and was just a luminous sphere, giving out light equally in all directions. Now imagine two large glass spheres, one with a radius of 1 metre and another with a radius of 10 metres, both centred on the light bulb.

If you sat anywhere on the one metre radius sphere (it's a very sturdy sphere) and looked towards the light bulb, you would see the same thing. No matter where on the 1 metre radius sphere you were, the light bulb would look as bright.

Now imagine that both large glass spheres were divided into 1 centimetre × 1 centimetre squares. If you remember back to school, the area of a sphere of radius *r* is equal to $4\pi r^2$. This means that the one metre radius sphere has an area of $4\pi \times 1 \times 1$, around 12.5 square metres or 125,000 square centimetres. The inner sphere therefore has 125,000 little squares on it. If you do the same thing for the larger sphere, it has 12,500,000 squares on it – 100 times more than the inner sphere.

Because the light intensity is constant – the light bulb pours out the same amount of light in every direction at every moment – the light intensity per square centimetre is 100 times lower in the outer sphere than the inner sphere. If you consider the intensity on a sphere 100 metres in radius, it would be 100×100, or 10,000 times lower. This is what is known as an inverse square law, where some quantity being measured decreases according to the square of the distance away from the source of whatever is being measured. Inverse square laws are crucial for our understanding of the universe. For one thing, they govern the intensity of light and other forms of electromagnetic radiation, which allow us to study what's going on out in space and show that the influence of these forces of nature never disappear, no matter how far away from the source of the force you are. Inverse square laws also apply to gravity – the force that controls the whole cosmos.

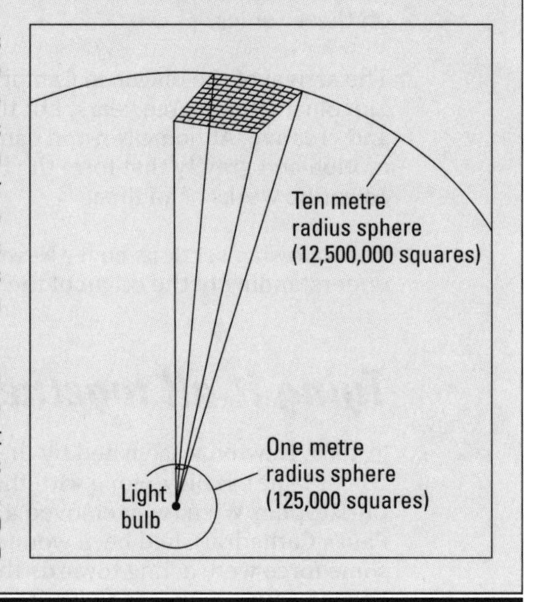

Newton's calculations revealed some monumental truths. For example, he showed that Kepler's third law – concering the relationship between a planet's year and its average distance from the Sun – can only be true if it involves some force that is directed towards the Sun and obeys an inverse square law. That force that is now known as *gravity*.

Newton's modesty?

The milled edge of the £2 coin contains part of Newton's most famous saying 'on the shoulders of giants'. The inscription honours Newton's scientific achievements but also recognises his position as warden of the Royal Mint beginning in 1696. While at the Mint, Newton introduced milled edges on British coinage in order to stop people clipping off the metal at the edges. You may be surprised to find out that Newton's knighthood came as a result of his services to the Mint rather than science.

About the reference to giants: The phrase comes from a letter sent by Newton to fellow physicist Robert Hooke discussing Hooke's work and how it expanded on the work of mathematician René Descartes. In it, Newton says: 'What Descartes did was a good step. You have added much in several ways, and especially in taking the colours of thin plates into philosophical consideration. If I have seen a little further it is by standing on the shoulders of giants.'

Many believe this phrase to be a sign of Newton's modesty, implying that he isn't a giant of scientific thought himself. However, others say that it's a hideous insult to Hooke, who was afflicted by a stoop. The jury is still out . . .

Newton's work is still the basis for our understanding of gravity's effects today. As he realised, gravity doesn't only work between planets and the Sun but is one of the universal forces of nature, an attraction that exists between all things. (However, as we explain in Chapter 4, a certain Albert Einstein showed that what we experience as gravity and what seems to be well described by Newton's ideas is really down to the shape of the universe.)

De motu didn't stop there. Newton also touched on the laws of motion and studied free fall and the trajectories of projectiles such as cannonballs, (following on from the work of Galileo). Newton's intention was to draw parallels between the forces that held the planets in their orbits and the forces that caused objects to accelerate in free fall. In short, he wanted nothing less than to make a universal law of gravity.

Appreciating Newton's Principia

De motu certainly caused a stir. Halley presented Newton's work to the Royal Society, an academy of the leading scientists of the day where the newest ideas were debated and adopted or discarded.

But Newton wasn't content with his first draft of *De motu* and revised it several times before finally publishing it in 1687 as *Philosophiae Naturalis Principa Mathematica* (*The Mathematical Principles of Natural Philosophy*) or *Principia* for short.

The following sections explore *Principia*'s, um, principal concepts and contributions.

Laws of motion

As well as the derivation of Kepler's laws, *Principia* contains what are now known as Newton's laws of motion:

- **Law of inertia.** An object at rest remains at rest unless acted upon by an external and unbalanced force. An object in motion remains in motion unless acted upon by an external force.
- **Law of acceleration.** The rate of change of momentum of a body is proportional to the resultant force acting on the body and is in the same direction.
- **Law of reciprocal actions.** All forces occur in pairs, and these two forces are equal in magnitude and opposite in direction.

Or in school physics terms:

- An object in motion remains in motion unless acted upon by another force.
- $F = m \times a$ (force equals mass times acceleration).
- For every action, there's an equal and opposite reaction.

Mass, momentum, and more

At the beginning of the *Principia*, Newton starts with some definitions. He invents the concept of *mass*, a measure of the amount of matter in an object. (By contrast, *weight* is a measure of the pull of gravity on an object.)

Nowadays, people measure both mass and weight in stones, pounds, or kilograms, depending on preference. But in scientific terms, the mass of an object is distinct from its weight – even though the two terms are sometimes used interchangeably.

To understand this subtle difference, consider the state of weightlessness experienced by astronauts. Although an astronaut in space has no weight (because of zero gravity), he or she still has the same amount of mass – 90 kilograms or whatever – as when on the surface of the Earth.

Newton also defined another concept – *momentum*, which equals the mass of an object times its velocity. *Velocity* is a measure of the rate at which an object changes its position. Velocity is subtly different from speed, which only measures how fast an object is moving.

Projectiles: Apples and cannonballs in the head

Schoolchildren are often taught that Newton formulated his ideas about gravity after an apple fell on his head. Although he may well have watched an apple fall from a branch during his time at the farm in Lincolnshire, no evidence exists to suggest he suffered an apple-induced bump on the top of his skull.

However, Newton's work with the laws of motion, the second in particular, made him wonder whether an object falling to the ground on the Earth is being attracted by the same force that keeps the Moon in its orbit.

Beginning with his earlier book *De motu*, Newton was also fascinated with projectiles. He started to think about what would happen if you fired a cannon horizontally off the top of a mountain. Of course, anyone who has ever fired a peashooter knows the answer: The projectile follows a curved path before eventually hitting the ground some distance away. The harder the cannonball is fired (or the pea is blown), the longer the curve.

Newton, however, pushed the idea to its limit, conceptualising powerful cannons that fired cannonballs farther and farther. He realised that at some point the curvature of the projectile's path would exactly match the curvature of the Earth – in effect the cannonball would be in orbit around the Earth and some time after firing your very powerful cannon, the cannonball would hit you in the back of the head. Figure 3-4 illustrates this idea.

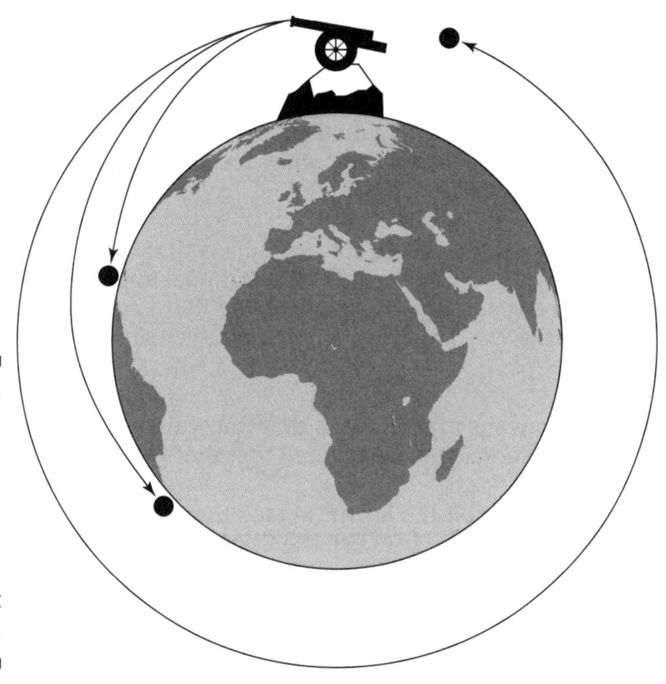

Figure 3-4: A cannon fires a cannonball from a mountaintop in Newton's thought experiment.

Furthermore, with a suitably high mountain (as high as the distance to the Moon) and a suitably large cannon (whose bore was the size of the Moon), Newton reckoned that you can put the Moon into the orbit that astronomers had known for centuries.

With this reasoning, Newton successfully linked the force acting on objects on the Earth's surface with the force holding the Moon in its orbit – and by extension, the planets in their orbits around the Sun.

A universal law of gravity

And Newton wasn't done yet! He suggested that gravity was everywhere, extending not only from the apple tree branch to the Earth's surface but also from the Moon to the Earth, from the planet to the Sun, and to the boundaries of the known universe.

Newton then combined his laws of motion and his mathematical derivations of Kepler's laws to work out the force between two objects – one of mass m_1 and another of mass m_2 – which were a distance r apart.

The size of this force is known today using the following formula:

$$F = \frac{Gm_1m_2}{r^2}$$

where F is the force and G is a constant equal to $6.67 \times 10^{-11}\,m^3\,kg^{-1}\,s^{-2}$, known as the *gravitational constant.* You'll notice that to calculate the force we need to divide the Gm_1m_2 part of the equation by the square of the distance between the two objects. This is because gravity obeys an inverse square law (described in the sidebar 'Inverse square laws').

One of the interesting things about this formula is that m_1 and m_2 are interchangeable. If you want to work out the gravitational force between the Earth and the Moon, you can set m_1 as the mass of the Earth and m_2 as the mass of the Moon or the other way around – no difference is made to the gravitational force.

What this situation means in practice is that the gravitational force that the Earth exerts on the Moon is the same force that the Moon exerts on the Earth.

So why does the Moon orbit the Earth? In fact, they're both in orbit around a common centre of gravity but because the Earth is significantly more massive than the Moon, the common centre is much closer to the Earth than the Moon. In fact, it is roughly 1,700 kilometres below the Earth's surface, so the Moon seems to orbit the Earth while the Earth just wobbles a bit around that point.

The formula also means that a falling apple is attracted by a gravitational force that is exactly the same as the force the apple exerts on the Earth. However, because the Earth is vastly more massive than the apple, the effect of the force on the Earth is imperceptible.

Believing that something gazillions of light years away (such as a distant galaxy, for example) has an effect on you may be difficult, but according to Newton's law of gravity, it does. In fact, if you know how massive something is and how far away it is, you can easily work out how much gravitational force it's exerting on you. The number may be vanishingly small but it does have an effect, according to Newton's reasoning.

This instantaneous action of gravitational force for a seemingly infinite number of distant objects is one reason why some believe that Newton's theory of gravity is all wrong. But that discussion must wait until Chapter 10.

Refusing to speculate on the cause of gravity

Newton did much to set in motion events that ultimately shook the foundations of the Church's view on the origins of the universe. But although *Principia* did much to explain the physical consequences of gravity, Newton wasn't prepared to speculate on its cause.

At the end of *Principia*, he wrote:

> *I have not as yet been able to deduce from phenomena the reason for these properties of gravity, and I do not 'feign' hypotheses. For whatever is not deduced from the phenomena must be called a hypothesis; and hypotheses, whether metaphysical or physical, or based on occult qualities, or mechanical, have no place in experimental philosophy.*

He concludes by saying: 'It is enough that gravity really exists and acts according to laws that we have set forth.'

Considering the impact of Newton's theories

Newton's work in the field that has become known as *classical mechanics* is both broad-reaching and revolutionary. *Principia* is quite possibly the greatest contribution to the advancement of scientific thought ever achieved – and that is without mentioning Newton's work in developing calculus, his invention of the reflecting telescope (which uses mirrors rather than lenses to magnify distant images), and his additional work on optics.

Most importantly, *Principia* provided a way for scientists to make predictions that can be checked and which had a basis in mathematical fact rather than metaphysical fantasy. Unsurprisingly, Newton's laws remained unchallenged for more than two centuries – and indeed are still in use today for most everyday situations.

Part II

Modern Cosmology: Going Off with a Bang

In this part . . .

It may be hard to believe but our greatest insights into the origins of the universe started off with scientists tinkering with electrical circuits and playing with magnets. Who could have guessed that the most famous equation of all time, Einstein's $E=mc^2$, would have taken shape from such beginnings? This part takes a look at that crazy-haired scientist's invaluable contributions to modern cosmology and how the universe would never seem the same again after he came along.

In this part we also explore the momentous event we've come to know as the Big Bang. But have we got it right? Most scientists believe in the Big Bang, but then most scientists believed in Newton's laws until Einstein came along. We end this part with a look at some of the alternative scientific theories for how the universe came to be.

Chapter 4

Bending the Universe: Magnets and Gravity

In This Chapter

- Discovering electromagnetism – and its impact on the universe
- Proving that the speed of light is constant
- Getting relative: Special and general relativity
- Bending space and time with Einstein

The most famous and popular physicist in history is without doubt Albert Einstein. The man who wrote the world's most famous equation, $E = mc^2$, really has no competition (though perhaps he wouldn't do so well in a beauty contest!).

But Einstein's fame is due to more than crazy hair and a three-letter formula. His work laid the foundations for a modern understanding of the universe, including the ideas that space and time are curved.

For more than two centuries, the ideas of Sir Isaac Newton had stood up to every conceivable test and seemed to represent some basic truth about the universe. Yet at the end of the 19th century, a growing number of astronomers and physicists began to realise that Newton's theories did not have all the answers.

So Einstein's universe-altering theories didn't arrive out of the blue. Here we discuss several other often unsung scientists who set the stage for Einstein's revolution through their work on the orbit of the planet Mercury, electromagnetism, and the speed of light.

In this chapter we talk about Einstein's theories of relativity and some mind-bending ideas about the shape of the universe. Don't worry – you don't need a PhD to follow these ideas, even though they sound daunting. The concepts are fundamental to our concept of how the universe began, so we've made sure you can understand them. In fact, you'll be able to amaze everyone else with your knowledge of some of the coolest stuff in cosmology.

Confirming Newton's Laws

Newton's well argued theories of motion and gravity (refer to Chapter 3) gained rapid acceptance in Britain and more widely over the half century after he formulated them, as other scientists and mathematicians, such as the brilliant Daniel Bernoulli, tested them and showed that they made accurate predictions.

The concept of gravity operating instantaneously and at a distance took longer for other scientists to accept. The French mathematician and philosopher René Descartes, for example, was a strong believer that forces could only operate through direct contact.

But what really convinced the sceptics that Newton was right was Edmond Halley's calculations of the orbit of a particularly bright comet and the prediction that it would return 76 years later. When the comet – later to be dubbed Halley's comet – appeared at the end of 1758, Newton's laws could no longer be ignored.

The perihelion of Mercury problem

In Chapter 3, we describe how Newton was unwilling to speculate on the 'how' of gravity when presenting his universal law of gravitation. However, a tiny departure from the expected behaviour of the planet Mercury came to prove that his universal law of gravitation wasn't universal at all.

Kepler's first law states that the orbit of the planets around the Sun is an ellipse of a fixed size. Yet observations of the planets show that although the size and shape of this elliptical orbit is the same for each year, the planet's path through space changes slightly from year to year, as shown in Figure 4-1.

The point of a planet's closest approach to the Sun is called the *perihelion* (as opposed to the farthest distance, or *aphelion*). If the universe consisted only of the planet and the Sun, apart from being a very dull place, the planet's orbit would stay the same every time around. Instead, because of the gravitational pull of the other planets, the point of perihelion moves, or *precesses,* in a circle around the Sun.

Newton predicted this precession, but although the gravitational pull of the other planets can account for most of this movement, a small but regular amount of precession remained unexplained. That is until Albert Einstein came onto the scene.

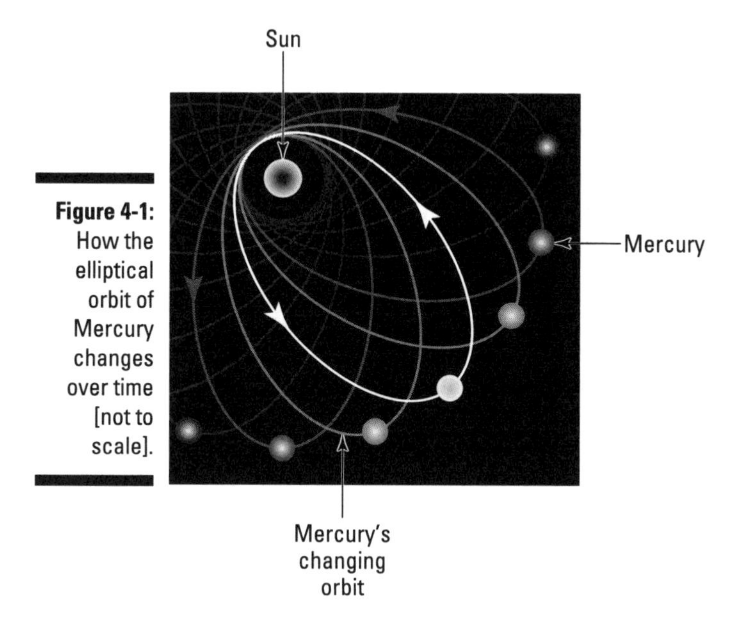

Figure 4-1: How the elliptical orbit of Mercury changes over time [not to scale].

Tripping the Light Electromagnetic: James Clerk Maxwell

Modern understanding of the universe really started to take shape around the end of the 19th century and the beginning of the 20th century.

One of the most significant scientists in the whole adventure was a Scottish physicist working at the University of Cambridge named James Clerk Maxwell (1831–1879).

The most significant of Maxwell's numerous contributions to science and mathematics concerned the close relationship between electricity and magnetism. He realised that these two phenomena were manifestations of the same thing, which he called the *electromagnetic field*.

Playing the electromagnetic field

The 'field' in the electromagnetic field is similar to the force fields so beloved of science fiction writers. Objects in sci-fi force fields are typically acted on by forces from a distance; objects in electromagnetic fields are affected by a force – electromagnetism – also at a distance.

Maxwell's theory of electromagnetic fields focuses on the space in the neighbourhood of electric or magnetic objects.

As you know from messing around with small magnets at home, each magnet has two distinct ends, normally called its north pole and its south pole. If you hold two magnets together with their north poles facing or their south poles facing, they repel each other. But if you put the south pole of one magnet close to the north pole of another, the magnets attract each other and stick together.

Furthermore, you've probably seen that when you scatter iron filings on a piece of paper and put a magnet underneath the paper, the filings arrange themselves in a series of curved lines. These *magnetic field lines* run from the magnet's north pole to its south pole, as illustrated in Figure 4-2.

But what do these field lines represent? In fact, they tell you about the electromagnetic force experienced by an object that's affected by the field. The direction of the lines show the *direction* in which the force acts, and the density of lines tells you about the *strength* of the force. At the poles of the magnet, the lines are more dense and this is where the force is strongest.

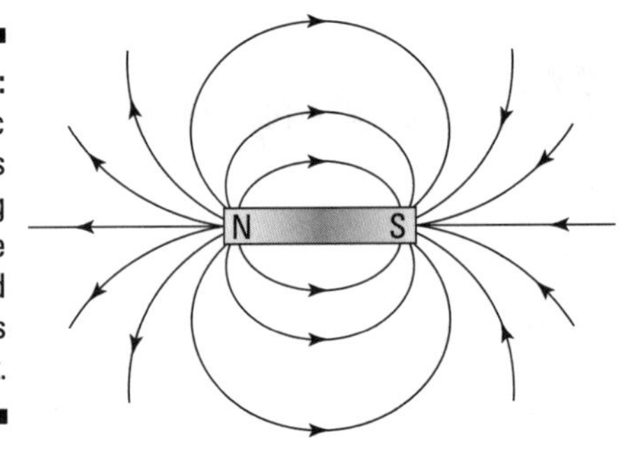

Figure 4-2: Magnetic field lines running between the north and south poles of a magnet.

It turns out that electrically charged objects generate fields too. Imagine two steel spheres, one with a negative electric charge and the other with a positive charge. You place positively charged ball bearings close to the two spheres and watch what happens. You can quickly measure the direction of the electric forces and their strengths. If you drew a diagram of the electrical force and its strengths you'd come up with something like Figure 4-3.

Figure 4-3 looks surprisingly like the magnetic field lines in Figure 4-2. Maxwell showed that they not only look the same, but they are the same, in certain circumstances. A clever bit of maths showed that when a magnet moves, it generates an electric field and when an electrical charge moves it creates a magnetic field.

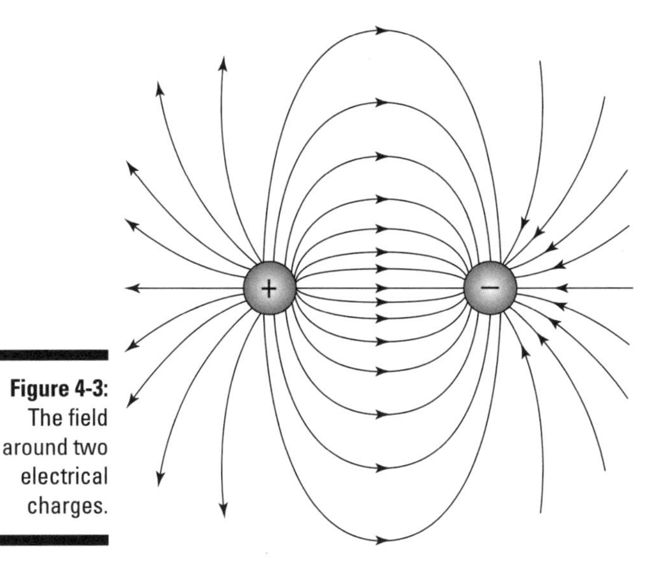

Figure 4-3: The field around two electrical charges.

You can see how the magnetic and electric fields are interrelated with a simple laboratory experiment comprising a wire connected to a switch and a battery. You also need a compass. When the switch is off, the compass points to the Earth's magnetic north. However, switch the current on and move the compass close to the wire, and you see that the compass needle is deflected. The moving electrical charge (the current) creates a magnetic field that shifts the compass needle.

Formulating wonderful equations

Maxwell's work unifying electricity and magnetism was the culmination of a lot of hard work by many scientists who had come before him. These physicists and mathematicians included André-Marie Ampère, best remembered today through the 13 amp (short for ampere) plug, and Michael Faraday who, among other things, came up with the reason why passengers don't get electrocuted when an aircraft is struck by lightning.

Maxwell used a series of eight mathematical equations to show how this relationship between electricity and magnetism works. In 1846, he presented his equations to the Royal Society. Some 40 years later his work was among the sparks that lit Einstein's genius fuse.

Realising a stunning coincidence

Perhaps the most remarkable thing that emerged from Maxwell's equations was a prediction that electric and magnetic forces travelled in wave-like oscillations. Not only that, but these oscillations travelled at a fixed speed.

Maxwell's calculations put the speed of magnetic waves at 288,000 kilometres per second, which is very close to the speed of light, which had already been measured by French scientist Hippolyte Fizeau in 1849 using a rotating mirror at 313,000 kilometres per second.

Maxwell realised the significance of that coincidence, and said as much in a scientific paper he published in the 1860s.

> *This velocity is so nearly that of light, that it seems we have strong reason to conclude that light itself (including radiant heat, and other radiations if any) is an electromagnetic disturbance in the form of waves propagated through the electromagnetic field according to electromagnetic laws.*

Maxwell's realisation was a triumph. He had deduced that light itself was an electromagnetic phenomenon. (Maxwell's theoretical equations were later proven by German physicist Heinrich Hertz, who generated radio waves, another form of electromagnetic radiation.)

This deduction was also the starting point for Albert Einstein's theory of relativity – Einstein was determined to find a way to make Maxwell's equations work whether you were stationary or moving. The key question is this: If you're sitting on top of a moving electrical charge (that is, you're at rest relative to that charge) will you still experience a magnetic field?

But his new insights into electromagnetism left Maxwell with a problem: What exactly was carrying around all these waves? His suggestion was that a substance called 'luminiferous ether' filled space, permeating everything, and carried electromagnetic waves the same way air carries sound waves. In fact, this idea had been around since the time of Aristotle (refer to Chapter 2) – the trouble was that no evidence of such ether had ever existed.

Getting Rid of the Ether: Michelson and Morley

Maxwell's idea that a mysterious substance called ether carried electromagnetic waves was definitely a problem for scientists until two US-based scientists came up with an experiment they hoped would measure the motion of the ether – if it existed – on the surface of the Earth.

Blowing in the wind

The experiments of Albert Abraham Michelson and Edward Williams Morley were based on the idea that as the Earth circled the Sun, the Earth was moving relative to the ether, creating a kind of 'ether wind'. That wind, the scientists thought, should cause slight variations in the observed speed of light that laboratory instruments can measure.

According to Dorothy Michelson Livingston (Michelson's daughter and biographer), Michelson explained ether wind to his own children by asking them to imagine a swimming race between two swimmers in a river with a strong current. Both swimmers must swim the same distance, and the following conditions apply:

- The first swimmer has to swim alongside the bank of the river. During the race, he spends half his time swimming with the current and half his time swimming against it.
- The second swimmer has to swim the same distance *across* the river and back again. During the race, he must take into account the push of the current at all times.

So who's most likely to win this imagined river race? Well, a bit of maths, which we don't go into here, shows that the person swimming across the river takes longer to swim the distance.

Michelson and Morley guessed that if an ether wind were blowing, it would act on light the same way that the current of the river acts on the swimmers. Scientists had considered this concept for some time. The trouble was that the difference in the measured speed of light would be quite small. Others had tried to make measurements, but the accuracy demanded was simply too great.

Testing the ether wind

The experiment to test the effects of ether wind was first performed by Michelson in Berlin in 1881 and then refined in 1887 by Michelson and Morley in the US. The intrepid duo set up an experiment that split a beam of light into two beams travelling at right angles to each other.

Specifically, the experiment involved directing a beam of coherent light (that is, of one specific wavelength) through a series of mirrors, as shown in Figure 4-4:

- First, Michelson and Morley aimed a beam of light through a half-transparent mirror, designed to reflect half of any light falling on it and allowing the rest to pass through the back.
- Given the angle and nature of the half-transparent mirror, half the light continued to move straight ahead, while the other half bounced off at a right angle.
- Each of the light beams then bounced off other mirrors and eventually returned to a light detector.
- In the light detector, the two beams created an *interference pattern,* which is a series of light and dark bands that form when the peaks and troughs of the two light waves re-combine.

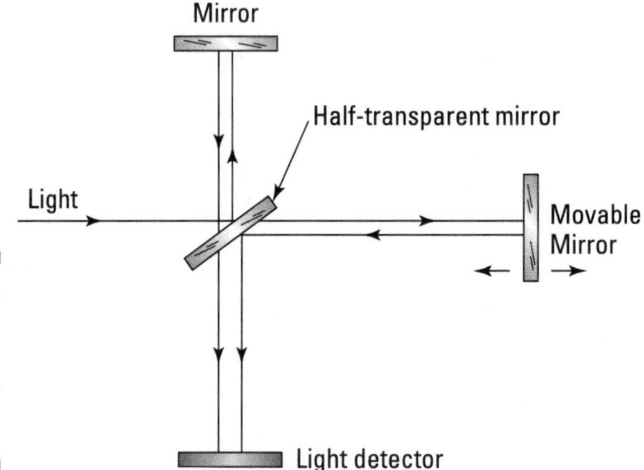

Figure 4-4: The Michelson-Morley experiment.

The equipment was mounted on a huge block of marble floating on a pool of mercury to avoid any vibrations that could mimic the sought-for effect.

Michelson and Morley hoped that the interference patterns formed by the two light beams would show that the beams travelled their reflected journeys at slightly different speeds – the speed of light plus the speed of the ether wind in one arrangement and the speed of light minus the speed of the ether wind when the contraption was rotated 180 degrees. But their results showed no differences between measurements taken at different orientations of the equipment or at different times of the day or year (when the rotation of the Earth on its axis and around the Sun should have had an effect) – a clear indication that the ether idea was nonsense. (In fact, scientists have repeated the experiment with increasing accuracy many times since Michelson and Morley, with the same findings.)

Response to Michelson and Morley's unsettling results

Many scientists and top thinkers of their day didn't welcome the results of the Michelson–Morley experiment. After the results were published, scientists tried to come up with reasons to explain away Michelson and Morley's findings.

Two of these scientists, George FitzGerald of Trinity College Dublin and Hendrick Lorentz from the University of Leyden in the Netherlands, independently came up with the same solution at roughly the same time – specifically that objects shrink as they move through the ether.

Lorentz and Fitzgerald suggested that the speed of light did in fact change relative to the movement of the ether, but that measuring that difference was impossible because the ether compressed rulers and other measuring devices in the direction of the ether wind by just the right amount to cancel it out. This idea may have been convenient, but it was a fairly ad hoc way out of the problems raised by Michelson and Morley.

However, the results of the Michelson–Morley experiments were deeply upsetting to scientists at the time. The findings suggested either that the Earth wasn't moving relative to the ether, or that the whole ether idea needed to be abandoned.

A few more years had to pass before the real implications of all this work came to light – thanks to the work of a young scientist named Albert Einstein. Michelson and Morley's experiment made the idea of the universally constant speed of light palatable and set Einstein on the path to his most famous equation, the one that shows the subtle interplay between mass and energy which governed the early universe.

Getting Relative with Albert Einstein

The most significant figure in 20th century science – in fact the most significant physicist since Sir Isaac Newton – is Albert Einstein. His work provided a foundation for today's understanding of the universe and led directly to the development of the Big Bang model (see Chapter 6).

Got the aptitude, but not the attitude

Einstein graduated from college with a degree in physics in 1900 but was unable to find a university post that allowed him to work toward his doctorate degree. Why? Well, Einstein had a problem with authority figures. The academics who taught him during his degree were upset by what they saw as his bad attitude – skipping classes, studying only what interested him, and so on. To get an academic post after graduation, Einstein needed a recommendation, something his professors were apparently unwilling to give.

Einstein's contributions to human understanding of the universe began in earnest in 1905, when he was 26 years old and working as a technical expert at the Swiss Federal Patent Office. As it turned out, his job at the patent office wasn't so bad. He later noted that the peace and quiet it offered gave him time to contemplate some of the problems of physics that had fascinated him since his teenage years.

Pondering a physical contradiction: Special relativity

Perhaps the most significant difficulties facing physics when Einstein was a lad were the contradictions between the two pillars of physics at the time: Newton's mechanics (refer to Chapter 3) and Maxwell's theory of electromagnetism (see the preceding section 'Tripping the Light Electromagnetic: James Clerk Maxwell').

Considering uniform motion

Newton's world view built upon some ideas laid out by the Italian scientist Galileo Galilei in the 17th century (who we talk about in Chapter 3).

According to Galileo, you cannot detect *uniform motion* (that is, motion along a straight line at constant speed) without an external reference point.

What does this mean exactly? Well, imagine yourself on a train moving at a constant speed along a straight track, as shown in Figure 4-5. If you throw a ball straight up in the air, it will come back down again to land in your hands (assuming you're co-ordinated enough to catch it!).

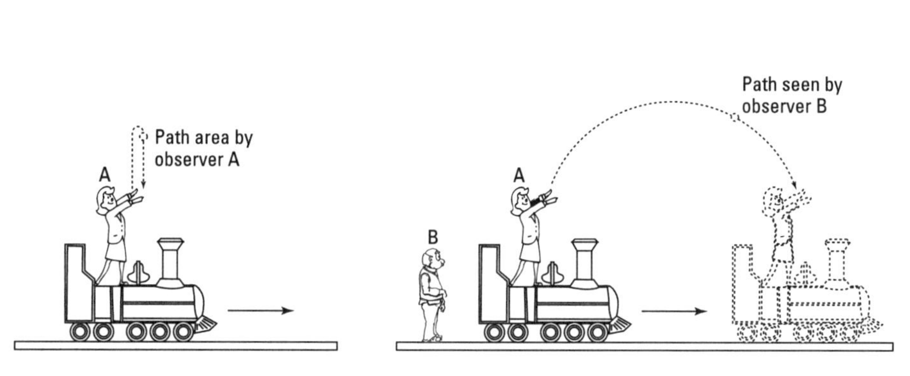

Figure 4-5: Catching a ball on a train. For the person on the train, the ball seems to travel straight up and down. But for an observer on the ground, it travels in an arc.

From your perspective on the train, the ball goes straight up and down – just as if the train were stationary.

However, the experience is different for a friend standing on the ground watching you as the train moves past. Before you throw the ball up, he can see it moving along at the speed of the train. From his perspective, when you release the ball into the air, the ball keeps moving *forward* as well as moving up, and then moving back down into your hands. The ball traces a kind of arc in the air, as Figure 4-5 shows.

Adding up speeds: Galilean relativity

To understand another key concept that Galileo proposed and Einstein pondered, again imagine you're on a train that is moving through the countryside at a constant 100 kilometres per hour. As you sit in your seat, a ticket inspector walks past from the front of the train towards the back, at a speed of exactly 5 kilometres per hour.

From your perspective – or *frame of reference* as physicists say – the inspector is moving away from you at a speed of 5 kilometres per hour. To keep pace with him, you'd need to walk at 5 kilometres per hour too.

But a friend on a platform outside the train has a much harder time keeping up with the inspector. She needs to travel at a different speed – 100 kilometres per hour minus 5 kilometres per hour, or 95 kilometres per hour – to keep pace with the inspector.

So what's the ticket inspector's real speed? Is it 5 kilometres per hour or 95? Well, Galileo said the answer depends on your reference frame.

Put another way, there's no way for you on the train to tell whether the ticket inspector is travelling at 5 kilometres per hour or 95. No method exists to distinguish between rest and uniform motion. Scientists refer to this experience as *Galilean relativity*.

For Galileo, and for Newton, the laws of physics were the same in all frames of reference in uniform motion.

Galilean relativity and electromagnetism

But Galilean relativity runs into a problem when you consider electromagnetic phenomena (like light), as Einstein realised.

In an attempt to reconcile the relativity of Galileo with Maxwell's laws of electromagnetism, Einstein worked on the basis of two principles, or postulates:

- The laws of physics are the same in all non-accelerated frames of reference (that is, ones that are in uniform motion, such as our train moving at a constant speed of 100 kilometres per hour).
- The speed of light is the same in all non-accelerated frames of reference.

At first glance, these two ideas seem contradictory, as Einstein himself pointed out in his 1920 book on relativity. He asked his readers to imagine a train (again!), moving at a fixed speed in a straight line. Additionally, Einstein asked readers to imagine the following:

- All the air has been sucked out of the scenario, and the train is moving in a vacuum.
- The speed of light travelling in a vacuum is constant, at a rate of about 300,000 kilometres per second. Einstein used the letter 'c' to represent the constant speed of light.
- A friend standing at a station platform sends a beam of light shining in the direction opposite to the movement of the train. (Pretend she's shining a torch – and wearing some kind of breathing apparatus so that she doesn't keel over in the vacuum!)

Considering Galilean relativity, as the preceding section 'Adding up speeds: Galilean relativity' describes, you may calculate that for the person sitting on the train, the speed of light seems to be c minus the speed of the train. But this can't be! If the laws of physics – and Maxwell's equations in particular – are the same in all frames of reference in uniform motion, the speed of light as seen from the train should be the same as the speed of light as seen from the platform. Something, somewhere has to give.

Stretching time

At first glance, the idea that light travels with the same speed relative to everything seems crazy.

For example, if you walk along a railway platform at 2 kilometres per hour and a train comes past you at 3 kilometres per hour in the same direction you're walking, the train's speed relative to you is 1 kilometre per hour. Surely, by the same logic, if you travel at 200,000 kilometres per hour, a light beam travelling parallel to you at 300,000 kilometres per hour seems to be going at 100,000 kilometres per hour. Right?

Einstein's answer was an emphatic 'no'. The speed of light is the same for all observers in uniform motion.

How can this be? Einstein's answer was that although the speed of light was constant, time wasn't. As he put it himself: 'My solution was really for the very concept of time.'

In June 1905, Einstein published the results of his solution in the journal *Annalen der Physik,* in a paper entitled 'On the Electrodynamics of Moving Bodies'. The implications were profound. Einstein's *theory of special relativity* – as it's now called to differentiate it from the general theory of relativity, which we explain later in this chapter – suggests that time *slows down* when you move. The theory also suggests that space contracts as you move.

In fact, the results of Einstein's calculations solved the problem exposed in the Michelson–Morley experiments (which we describe in the earlier section 'Getting Rid of the Ether: Michelson and Morley').

What Einstein was able to show was that his view of relativity meant that the length of something (a train for example) contracted by a factor of

$$\sqrt{1 - \frac{v^2}{c^2}}$$

where v is the relative velocity involved and c is the absolute speed of light. This equation also shows why the speed of light is considered an absolute limit – if v goes above c then the amount under the square root sign becomes negative. Try doing that on your calculator!

Scientists George FitzGerald and Hendrik Lorentz had previously suggested that because objects are made up of atoms and molecules held together by the electromagnetic force, and because motion had been shown to deform electromagnetic fields, objects should get smaller when they moved. Einstein showed that this contraction was a direct consequence of relativity.

An extraordinarily good year

For Einstein, 1905 was a very good year indeed, because he published four scientific papers that transformed science, despite only having a job as a patent clerk.

- His first big paper of 1905 was on the nature of light and suggested that light interacts with matter as discrete 'packets' of energy.
- His second paper explained the random movement of very small objects, called *Brownian motion,* as direct evidence of molecular action. This concept provided support for the existence of the atom.
- His third major contribution came in a paper on the electrodynamics of moving bodies, which proposed the special theory of relativity.
- His fourth paper on the equivalence of matter and energy showed that special relativity led to the most famous scientific formula of all: $E = mc^2$.

In fact, these papers were only *some* that he published in 1905 – a couple of others are important as well, but less dramatic. Altogether 1905 added up to what has been called Einstein's *Annus Mirabilis,* his extraordinary year.

To get a sense of how the absolute speed of light (that is, fixed in whatever frame of reference you consider) causes time to slow down (or *dilate* in scientific jargon), imagine that you're on a moving train – yet again! You're shining a laser light between two mirrors, one on the floor of the carriage and one on the ceiling, as Figure 4-6 shows.

Imagine that you can measure the length of time required for the light to bounce up and down from floor to ceiling and back again. Also imagine that the light bounces up and down between the mirrors exactly 20 million times every second.

Now imagine that a friend standing on a railway platform can also see the light bounce up and down. She sees it move up at a diagonal, and then back down again at a diagonal. From her perspective, the light travels farther than it does from your perspective. But remember that according to Einstein, both you and your friend measure the same speed for the light.

The upshot of all this is that if your friend on the platform measures the time it takes for the light to bounce 20 million times, she would find that it takes longer than a second as measured by the watch on her wrist. Thus, from her perspective, time is moving more slowly on your train than for her on the platform.

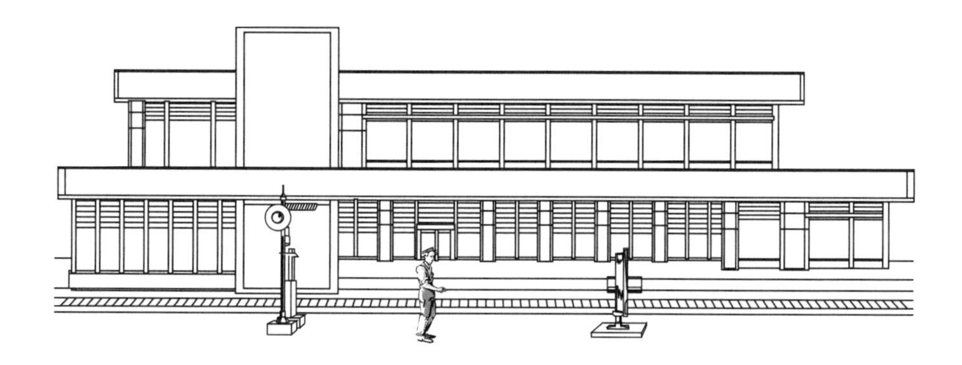

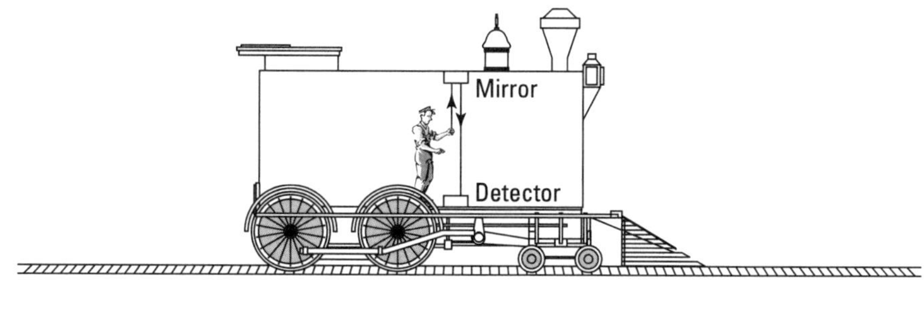

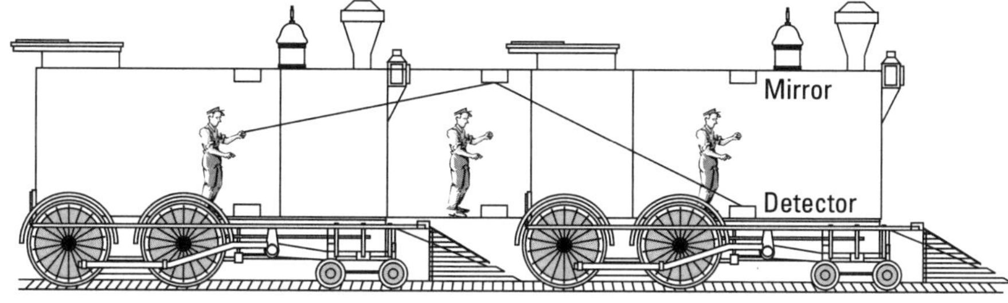

Figure 4-6: (Top) You see the light beam go up, bounce off, and come straight back down. (Bottom) Your friend sees the beam go up along a diagonal, bounce off, and move down along another diagonal.

This ramification is one of the most important of special relativity: Time in a moving frame runs more slowly. By extension (because the speed of light, a distance divided by a time, has to remain constant) Einstein also showed that space is shortened in a moving frame.

Suddenly, time and space were no longer the fixed background to the goings on of the universe as they were once considered. Other scientists had come close to unveiling these truths about the universe, including the French mathematician Henri Poincaré, but Einstein is the one who took the final step.

These changes in time and size probably seem odd and don't match with your everyday experience. Why? Simply because you never travel at anything like the speed of light. But scientific experiments have shown over and over that these changes really do happen.

Connecting mass and energy

Soon after Einstein published his special theory of relativity in June 1905, he followed it up with another brief scientific paper titled 'Does the Inertia of a Body Depend on Its Energy Content?'. In the paper, he used the principles of special relativity to show that when an atom emits light or other electromagnetic radiation, such as a laser or a radioactive source giving off X-rays, its mass decreases.

Mass is similar to, but different from, weight (Chapter 3 has a fuller discussion).

In a nutshell, Einstein's paper showed that the mass of a body is a measure of how much energy it contains. That is, matter and energy are interchangeable!

Interestingly, the paper in which Einstein formulated this idea doesn't contain the famous formula $E = mc^2$ anywhere. Rather, it says: 'If a body gives off the energy L in the form of radiation, its mass diminishes by L/V^2.'

Breaking down this sentence into its parts shows how the sentence turns into the well-known equation:

- Einstein used the symbol V to represent the speed of light in his paper. Nowadays, scientists use the letter 'c'.
- L represented energy, which is normally written as E.
- L/V^2 means L divided by V^2, where V^2 means V squared, which is another way of writing V multiplied by V. Rewriting this equation in a mathematical way, you get the following:

  Mass = E divided by c^2

  From there, simply multiply both sides of the equation by c^2 to get, wait for it (drum roll, please. . .):

  $E = mc^2$

And what does it all mean? A lot. Specifically:

- **Energy is a form of mass, and mass is a form of energy.** Because the factor that relates them is an enormously large number (the speed of light squared), even the tiniest fragment of matter has the potential to release an enormous amount of energy.

- **An object's mass increases as its velocity increases.** As an object gets close to the speed of light, its mass increases in the direction of infinity to the point that no amount of effort can speed it up any further. That's why nothing with mass can possibly travel faster than light.

Moving on to general relativity

Einstein knew that his work on special relativity was important, but he also realised from the start that his special theory of relativity was restricted to a particular kind of motion – motion without acceleration.

Finding a way to formulate a more general form of relativity, particularly to extend it to include gravity, proved enormously tricky. He spent more than a decade working on it, before finally getting there in 1916.

Making gravity and acceleration interchangeable

In 1907, Einstein had what he later called 'the happiest thought of his life', which concerned a painter falling from the roof of a house.

Now, Einstein was no sadist! His happiness didn't come from the poor painter's agony but from something the painter said afterwards. The painter explained that he had felt nothing until he hit the ground. Einstein realised this meant that the painter wasn't aware of the pull of gravity while he was falling.

You can think about this another way. Imagine that you're the 17th-century physicist Galileo Galilei. You're standing at the top of the Leaning Tower of Pisa about to perform your most famous experiment by dropping a cannon-ball and a wooden ball from the top to see whether one falls faster than the other. (Yes, Galileo is famous for this experiment even though he probably didn't actually perform it – as we mention in Chapter 3. Still, it illustrates a point nicely!)

But just as you're about to drop the two balls, you lose your footing and fall off the tower. As you're falling, being the dedicated scientist you are, you decide to perform the experiment anyway and release the balls. To your surprise, the balls don't fall away from you. It's as if gravity doesn't exist.

Einstein's inspired conclusion based on a similar thought experiment was to realise that acceleration and gravity are equivalent. The falling person – whether a painter or Galileo – doesn't know whether he's being pulled by gravity or being accelerated at the same rate in some other way – apart from having just fallen off a tall building, of course.

This *equivalence principle,* which says that gravity and acceleration are indistinguishable, formed the cornerstone for Einstein's *general theory of relativity.*

The equivalence principle and light

Einstein realised that the fact that gravity and acceleration are interchangeable has a profound impact on light.

To see why, imagine yourself on a spaceship accelerating through space. The side of the craft has a small hole that lets in a beam of light that shines across the cabin and hits the wall on the other side, as Figure 4-7 shows.

Because the spaceship is accelerating, when you measure where the light hits the wall, the light is slightly lower than if the ship had been standing still – as if the light has bent downward in an arc toward the floor.

That's all pretty easy to imagine, you may think. But now consider that gravity and acceleration have the same effect. So gravity should have the same impact on the light shining into the spaceship – it should bend light a little.

Figure 4-7: A light beam appears bent to the scientists in the accelerating spaceship.

Bending space and time

As Einstein contemplated the idea of light being bent by gravity, he realised that he needed to use another unusual form of geometry to help describe it. (See the sidebar 'Ripping the timespace continuum' for yet another form of geometry that was invaluable to Einstein's work.) Talking to mathematicians Einstein came across the idea of *Riemannian geometry,* which works on curved surfaces rather than flat planes.

After years of hard work with complex mathematics, Einstein emerged victorious with his *general theory of relativity,* in which the laws of physics are valid in any reference frame in any kind of motion, not just non-accelerated frames as in the special theory (as we describe in the preceding section 'Pondering a physical contradiction: Special relativity').

In fact, general relativity showed that gravity wasn't causing light to curve. Instead, *spacetime* – a mathematical model that combines space and time into a single construct, where space is usually three-dimensional and time plays the role of the fourth dimension – itself is curved and the light is simply taking the shortest path to its destination.

What, you may ask, does that mean? Good question. Perhaps the best way to envisage spacetime is to picture it as a sheet of rubber, stretched horizontally, as in Figure 4-8. (Of course, we must condense the four dimensions of spacetime into just two to make it easier to imagine – as well as publish on a two-dimensional page!)

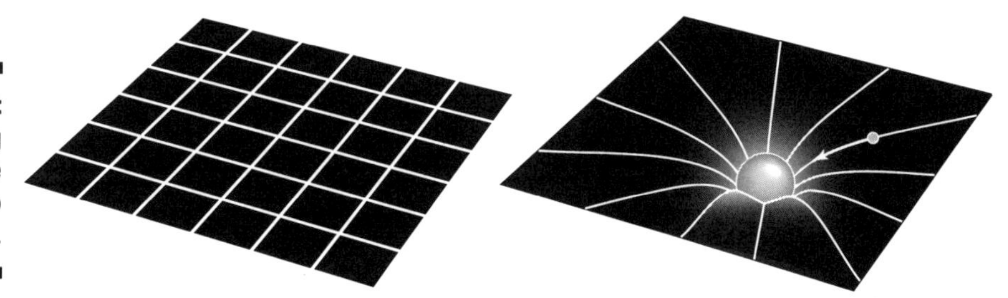

Figure 4-8: Envisioning spacetime in two dimensions.

If you roll a nice spherical grape across the rubber sheet of spacetime, it rolls in a straight line – just as you expect light would, or a comet streaking across space.

Now imagine putting a heavy watermelon, for example, in the middle of the sheet. What happens? It makes the rubber curve and stretch down toward it, forming a dip. If now you place the grape on the rubber sheet, it rolls down toward the watermelon. If you try to roll the grape past the side of the melon, the grape takes a curved path.

What Einstein's general theory of relativity shows is that what we assume to be gravitational forces – the pull of an apple to the ground, the force keeping the Moon in orbit around the Earth and the planets around the Sun – are not actually forces in the widely understood sense after all. Instead, the curving effect of mass on the rubber sheet of spacetime makes it seem as if forces are acting.

Imagine that the watermelon is the Sun and the grape is the Earth and start spinning the grape around the rubber sheet at just the right speed. If no friction came from the sheet, the grape would just keep whizzing around the watermelon in an elliptical orbit. No force is causing this to happen, merely the curvature of the sheet.

Proving spacetime is curved

Calculating that spacetime is curved is one thing, but proving it is quite another. The problem is that even on large scales the curvature of timespace is incredibly gentle. Einstein himself recognised the problem, saying that the curvature was 'exceedingly small for the gravitational fields at our disposal in practice'.

So here's how astronomers test Einstein's contention that spacetime is curved:

- **Take one very heavy heavenly body.** The heaviest thing on hand that is convenient for humans to observe is Earth's own star, the Sun.
- **Wait for a solar eclipse.** To test the theory that space and time are curved, you have to look at light passing close to a heavy body because then you can see that body's gravitational pull bending the light. Normally the Sun is so bright that other light is drowned out. (Have you ever seen stars at midday?) Luckily, nature provides an excellent, occasional sunshade – the total solar eclipse.
- **Look at stars close to the Sun's edge.** If Einstein is right, when you look at a star close to the edge of the Sun during an eclipse, its apparent position shifts because the light travelling from the star is warped by the gravitational pull of the Sun.

  Einstein reckoned that starlight grazing the edge of the Sun would be curved by a minute fraction of a degree equal to about the size of a penny seen from a distance of 1.5 miles.
- **Get out your star atlas.** Compare the position of a recognisable star on the day of the eclipse with its expected position.

  During the total solar eclipse of 29 May 1919, astronomer Arthur Eddington visited the island of Principe off west Africa and took a photograph of stars in the Hyades cluster, which at that time were close to the edge of the Sun. Despite almost being unable to make an observation because of cloud cover, the skies cleared enough for Eddington to see and record that the star had in fact appeared to shift its position enough to confirm Einstein's ideas.

Ripping the spacetime continuum

Fans of science fiction potboilers are familiar with the word *spacetime,* which crops up in regularly rolled out clichés, such as 'a rip in the fabric of the spacetime continuum'. But what exactly is spacetime? Unsurprisingly, it's a combination of space and, you guessed it, time.

Ever since the time of the Greek mathematician Euclid, humans have considered the universe around them to be three dimensional. Imagine, for the sake of argument, that the Earth is the centre of that universe and that you call the direction towards the Sun the x-axis. If you then call the direction towards the centre of the Milky Way galaxy the y-axis, you can then find a third direction that's at right angles to both of these, which you can call the z-axis, as the figure shows.

You can now define any point in the universe by three coordinates, x, y, and z, which represent a distance along each of the x, y, and z axes from the Earth. In this geometry – known as Euclidean geometry – one of the key points is that lines that are parallel always remain so no matter how far you travel through the universe.

But some mathematicians in the 19th century began to realise that other possible geometries exist. One that proved to be invaluable for Einstein was a four-dimensional geometry, conceived by Hermann Minkowski, one of Einstein's teachers. This geometry included the x, y, and z dimensions of Euclid – and added a fourth dimension, t, for time. Now, rather than a point in the universe defined by the coordinates x, y, and z, you consider events defined by the four coordinates x, y, z, and t.

Minkowski unveiled this concept in 1908, after realising that Einstein's special relativity can be conveniently handled in this way. (We describe special relativity in the section 'Pondering a physical contradiction: Special relativity'.)

Minkowski said at the time: 'Space by itself, and time by itself, are doomed to fade away into mere shadows, and only a kind of union of the two will preserve an independent reality.' Thus was born the concept of spacetime.

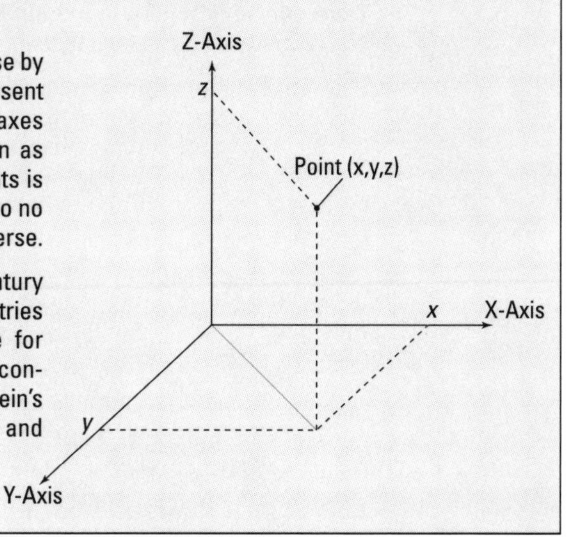

The gravity of a very heavy object acts as though it's bending light, just like a lens does. (In fact, the concept has become known as *gravitational lensing.*) Astronomers have since observed some spectacular effects of the gravitational lensing power of clusters of galaxies. Because of their heavy concentration of mass, galaxy clusters distort light vastly more than a single star like the Sun. In fact, some clusters of galaxies are surrounded by weird bright curves. Astronomers now believe that this light is from objects behind the clusters, which have had their light massively distorted, in an effect similar to that you see when you look at a digital alarm clock through a full glass of water.

As scientists after Einstein began to explore the universe more thoroughly using high powered telescopes, they discovered that the light from distant stars is also an important tool for measuring the expansion of the universe. Yes, although the universe is immense, it continues to expand. The universe's expansion tells us much more about its origins and its future (see Chapter 5).

Chapter 5

Measuring the Universe

In This Chapter

- Classifying stars by magnitude and more
- Measuring distances between stars
- Studying variable stars and nebulae to expand understanding of the universe
- Putting the Milky Way in perspective with Edwin Hubble

Throughout history, philosophers and scientists have tried to estimate the size of the universe. Why? Certainly, we humans have an innate desire to know our place in the universe, however inconsequential that place turns out to be. But anyone interested in the origins of the universe must know how big it is before fully starting to comprehend how and where it began.

Cosmology – the science of the universe – is quite unlike other sciences, such as biology and chemistry. Biologists can grasp specimens in their hands in order to dissect them, and chemists are never more at home than when holding pipettes and test tubes. Cosmologists, by contrast, can never hold a single star – let alone the entire universe – in their hands to experiment with. Indeed, the farthest humans have ever reached with the aid of human-made scientific instruments is just beyond the edge of the solar system. Yet astronomers, exercising considerable intellect and ingenuity, have discovered and developed amazing ways to find out more about the stars and other objects in the heavens.

This chapter looks at how, in the absence of being able to grasp the objects of curiosity in their hands, astronomers established the chemical make-up of distant objects, determined how fast they are travelling, and perhaps most important of all, figured out how far away these objects are.

Examining All Those Twinkling Little Stars

When you look up into the night sky with the naked eye, three things are instantly evident:

- **Stars aren't uniformly distributed across the sky.** In some places, stars appear to be clumped together, and in others, individual stars look as though they are on their own. In some places, very few stars are apparent at all.
- **Stars appear to be tiny, bright points.** To the unaided eye, no stars (other than the Sun) look the size of a basketball – or any other sort of ball for that matter. Stars are all so incredibly distant that you can't visibly see their actual shapes.
- **Stars vary in their brightness.** Some are so faint they're only visible when you're out in the country, away from streetlamps. Others are bright enough to shine in the night sky above Piccadilly Circus (on those rare occasions when London isn't blanketed by cloud).

Quite a leap of the imagination (or even faith if we're honest) was necessary for the first astronomers to understand that the Sun was the same type of object as the distant stars.

The following sections discuss science's gradual observation, categorisation, and explanation of the vast night sky.

Looking more closely

To more fully understand the universe, you must use instruments a little stronger than your eyes.

For example, with just a pair of binoculars or a small telescope, you can look at the Moon and see much more detail – craters upon craters upon craters in fact.

Or you can turn these simple instruments towards a planet. Many of the brightest objects in the sky after the Sun and the Moon are planets. To the uninitiated they look like stars – bright dots in the sky – but their apparent movement across the sky relative to the seemingly unchanging background of stars (other than the daily rotation of the Earth) marks them out as something different. The view through binoculars or a telescope confirms this difference. The bright dots turn into discs, and some of them, notably Jupiter and Saturn, even have moons of their very own.

However, when viewed through binoculars or a basic telescope, the stars remain as resolutely pointlike as when viewed with the naked eye. What's more, although you can see more stars with binoculars – millions compared with just a few thousand with the naked eye on a very clear night away from artificial light (perhaps a hundred or so in the middle of a city) – their distribution remains similar. The heavens contain patches where stars seem clumped together (either in loose associations or more obvious clusters) while other areas have few stars.

Perhaps the most notable thing about viewing the sky with something other than the naked eye is the appearance of the Earth's galaxy, the Milky Way. The faint veil that slashes across the sky to the unaided eye is actually made up of stars – and the more powerful your instrument, the more stars you can see.

A sweep of the sky with a pair of binoculars or a telescope also throws up some oddities, such as variable stars and fuzzy blobs know as *nebulae,* which we cover in greater detail in the later section in this chapter 'Probing the mysteries of nebulae'.

Comparing the colour of stars

On first glance, stars all look white, but this observation is not actually true. Study the sky for a while, even with your eyes alone, and you start to notice subtle shades in the colour of light produced by stars.

Consider the constellation of Orion the Hunter, well known to the people of the northern hemisphere. This constellation's most conspicuous feature is the hunter's 'belt', a row of three bright stars from which a hunting knife or sword (made up of stars) hangs. With a healthy dose of imagination, you can even make out the shape of a hunter.

The bright star at the shoulder of Orion is called Betelgeuse (or sometimes Alpha Orionis). If you compare Betelgeuse with neighbouring stars, you can see that it has a distinct red hue. By contrast, switch your view to the foot of the constellation, to the star Rigel (also known as Beta Orionis), and you can see a much bluer look to it. (And you've surely noticed that the Sun has a rather yellowy tinge to its light!)

Electromagnetic spectra

Visible light is a form of *electromagnetic radiation,* which travels through space as waves. As Figure 5-1 shows, electromagnetic radiation includes radio waves, microwaves, X-rays, and infrared and ultraviolet light, in addition to visible light. The distance between successive crests in these waves is known as the *wavelength.* What distinguishes the different forms of radiation from one another is their wavelength.

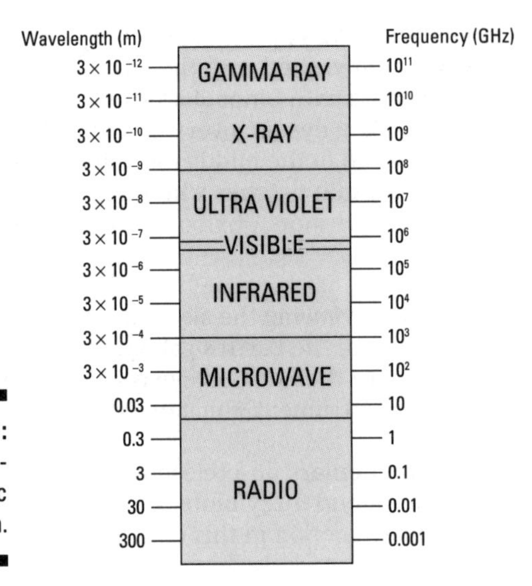

Figure 5-1: The electromagnetic spectrum.

Visible light, for example, has wavelengths of around 10^{-6} metres (a thousandth of a millimetre), with different colours of light having very slightly different wavelengths. X-rays have much smaller wavelengths, of around 10^{-9} metres, whereas radio waves have longer wavelengths, ranging anywhere from 10 centimetres to hundreds of kilometres.

When you look at a source of light – or indeed any other form of electromagnetic radiation – you're actually seeing a wide range of different wavelengths all mingled together (unless the light is a very pure source like a laser). If you analyse the light coming from a standard light bulb, for example, you see that it's made up of a wide spectrum of light of all the different colours of the rainbow. Use the right equipment and you'd also find that a light bulb emits radiation at wavelengths above and below the spectrum of visible light.

Electromagnetic radiation and stars

Since the middle of the 19th century, astronomers have used similar methods to analyse the electromagnetic radiation being transmitted from stars. Figure 5-2 shows one such spectrum.

For any object that emits radiation, such as light or X-rays, the wavelengths of the radiation that emerge from it are typically governed by the object's surface temperature. That's why things glow 'red' hot and 'white' hot at different temperatures.

The troughs in Figure 5-2 are called *absorption lines.* Absorption lines occur because certain chemical elements and molecules present in stars absorb electromagnetic radiation of specific wavelengths. For example, hydrogen, the most abundant element in stars, absorbs electromagnetic radiation with wavelengths of 410, 434, 486, and 656 nanometres (10^{-9} m). Take a look at Chapters 9 and 12 for more on the chemical composition of stars.

Stellar spectra exhibit thousands of such lines, not just from the presence of hydrogen, but also from other common elements such as helium, calcium, iron, sodium, and even unusual molecules such as titanium dioxide.

This information is all very interesting but what use is it? Well, absorption lines enable you to figure out what a star is made of without having to visit it. The depth of the troughs or the relative darkness of the absorption lines can tell you how much of each element a star contains – but only after you take into account that some elements absorb and re-emit radiation more efficiently than others.

Absorption lines also helped astronomers in an unexpected way. When the spectra of stars are compared with those produced by the Earth's closest star, the Sun, an oddity emerges. Although many stellar spectra showed similar collections of lines – as you may expect if all stars are made of similar materials – some had deeper and wider troughs than others. Even more curious, the absorption lines in some spectra were shifted along the spectrum to higher or lower wavelengths. The explanation for this shift is a key step in understanding the nature of the universe, as the later section 'Mixing it up with variable stars' explains.

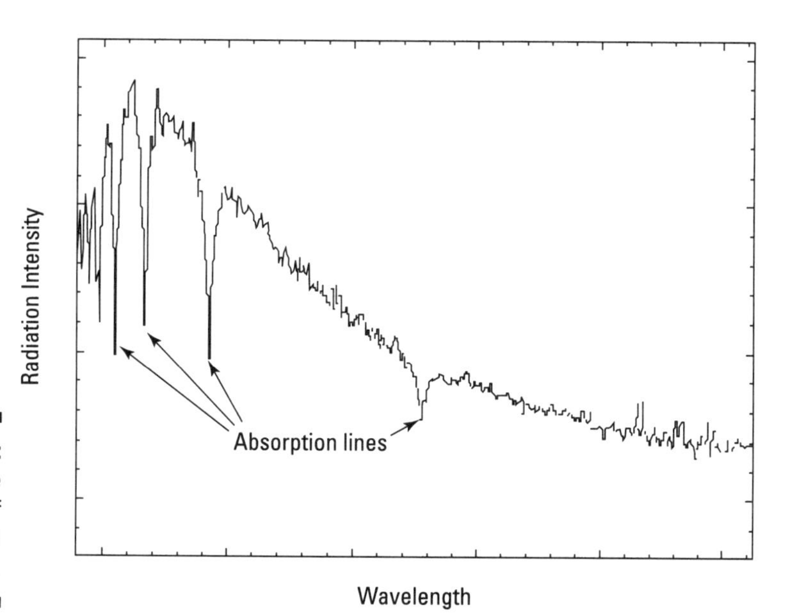

Figure 5-2: The spectrum of a typical star.

Classifying the Stars

When astronomers began to realise that different stars didn't have exactly the same composition, they began to think about classifying them in some way. The system of classification used today came about in the 1890s as a result of work by a team of scientists at Harvard College Observatory.

Organising by surface temperature

Edward C. Pickering, Annie Cannon, Antonia Maury, and Williamina Fleming of Harvard College Observatory classified thousands of stars according to the strength of the absorption lines of hydrogen in their spectra (see the earlier section 'Electromagnetic radiation and stars'). Those stars with the most hydrogen were classed as A, the next B, and so on. In all, these astronomers observed 22 distinct classes.

However, this system has since been refined and many of the 22 classes dropped. Unfortunately for people who like order in their lives, the system was also reorganised to put the stars in order of their average surface temperature, which meant that O and B were suddenly in the wrong alphabetical order. As a result, stars are now classified as follows:

- **O class stars** have surface temperatures of 30,000 kelvin and above (see Appendix for discussion of the Kelvin). They are typically blue in appearance and emit lots of ultraviolet light. Their spectra reveal lots of absorption lines of helium. One such star is Delta Orionis, one of the stars in the belt of the constellation Orion.
- **B class stars** have surface temperatures that range from around 10,000 to 30,000 kelvin. The spectra of these blue-white stars reveal strong helium absorption lines, as well as lines from silicon, nitrogen, and oxygen. Rigel, the brightest star in Orion, is a B class star.
- **A class stars** have surface temperatures that range from 7,500 to 10,000 kelvin and a white colour with perhaps the slightest hint of blue. Sirius, the brightest star in the Earth's sky, is a class A star. Class A stella spectra have the strongest hydrogen absorption lines.
- **F class stars**, such as Canopus (the second brightest star in the night sky), have surface temperatures of 6,000 to 7,500 kelvin. As soon as a hint of yellow enters a star's colour, you're out of the A class and into the F class of stars. These stars show lines of hydrogen, calcium, and other heavier elements.
- **G class stars,** such as the Earth's Sun, have surface temperatures in the 5,000 to 6,000 kelvin range and a yellow colour. With G stars, absorption lines from heavier elements become stronger.

- **K class stars** have surface temperatures of 3,500 to 5,000 kelvin and are the second most common type of star. Around one in six stars in the visible universe is a reddish orange K class star. Alpha Centauri B is a K class star. The lines of hydrogen are very weak in K class spectra whereas metals become increasingly dominant.
- **M class stars** have surface temperatures of between 2,000 and 3,500 kelvin. Red M class stars are the most common stars you can see. Most are small, relatively cool stars that astronomers call red dwarfs, such as Proxima Centauri, the closest star to the Earth other than the Sun. Their spectra show absorption lines caused by the presence of molecules such as titanium oxide.

Many astronomers first memorise the letter sequence of the revised star classes by using the mnemonic 'Oh be a fine girl/guy – kiss me', although remembering the temperature sequence is a lot harder.

The preceding temperature-based letter classes are further subdivided into nine numbered subclasses on the basis of their temperature. The Sun is a G2 star, for example. But temperature-based star classes don't tell you everything. Astronomers are also interested in the brightness and size of stars.

Differing magnitudes

Ejnar Hertzsprung was a Danish astronomer who did his ground-breaking work on star classification around the turn of the 20th century in Copenhagen. In 1905, he published a paper called 'Zur Strahlung der Sterne' ('On the Radiation of Stars'). Hertzsprung realised that some stars with the same class, and therefore the same surface temperature, had very different brightnesses, or in astronomer-speak, *magnitude* (see sidebar Star light, star bright on page 81).

Two types of magnitude are in regular use – absolute and apparent. *Absolute magnitude* refers to the intrinsic brightness of an object whereas *apparent magnitude* is how bright the object appears to be from where you observe it on the Earth, which depends on the distance to the object.

If the difference between absolute and apparent magnitudes seems hard to grasp, think about a handheld torch. The intrinsic brightness of the beam, which depends on the light bulb and the amount of power available from the batteries, doesn't change. However, you perceive a difference in how bright the torch's light appears when you look at it from a few feet away or a few miles away.

The star Rigel in the constellation Orion is a class B star and so is a star called Spica, which is the brightest star in the constellation of Virgo. Yet these two stars have very different brightnesses: Rigel's intrinsic brightness is many times greater than Spica's, despite being in same class.

Henry Norris Russell carried out similar classification work independently of Hertzsprung. Both scientists plotted graphs of the intrinsic luminosity of stars against their temperatures (*luminosity* is the rate at which a star emits energy). The resulting diagrams are now known as Hertzsprung-Russell diagrams.

If you plot a good sample of stars on a Hertzsprung-Russell diagram, as in Figure 5-3, you eventually see that stars aren't randomly distributed about the diagram but concentrated in several areas. The bulk of the stars, around 90 per cent of them, lie in a diagonal band stretching from the top left to bottom right. These stars are known as *main sequence stars.*

- The group of stars in the bottom left of a Hertzsprung-Russell diagram, with higher temperatures and lower luminosity, are known as the *white dwarfs*.
- The group of stars in the top right of a Hertzsprung-Russell diagram, which are very luminous but not all that hot, are known as the *giants* and *supergiants* because of their large size.

A Hertzsprung-Russell diagram can show how stars develop through their lives. Stars spend most of their lives burning hydrogen, which is the reason why the main sequence is so well populated. Stars move slightly along the main sequence as they age but when the nuclear fusion of hydrogen ends (as we discuss in Chapter 12), they turn off the main sequence. Our good old Sun currently lies close to the middle of the main sequence and will eventually turn off the main sequence, heading to the top and right of the diagram as it becomes a red giant.

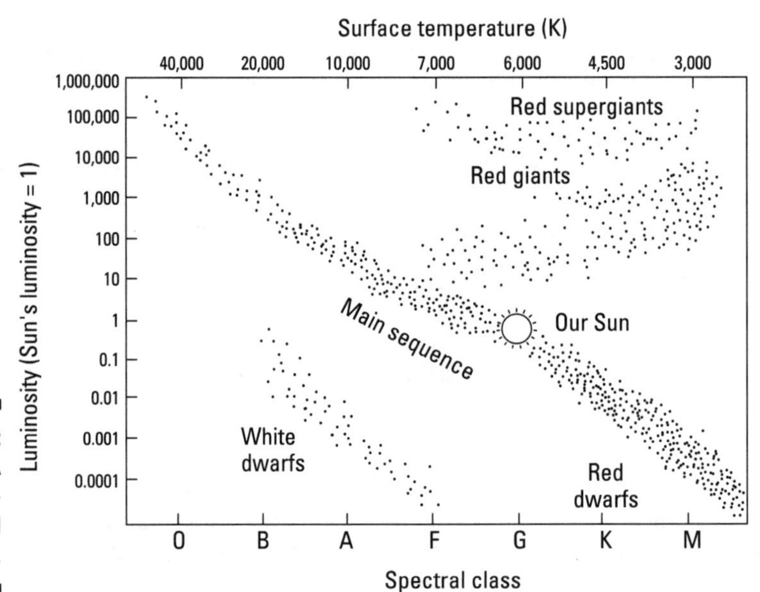

Figure 5-3: A Hertzsprung-Russell diagram.

Star light, star bright: Magnitude through the ages

When a young star gazer decides to become an astronomer, he or she must become familiar with a huge amount of specialised terminology. Like other scientists, astronomers use jargon to make sure that no one else can do their fun jobs without studying really, really hard! One of the first essential bits of jargon is that you never refer to a star's brightness, but rather to its *magnitude.*

Using magnitude – a word that in regular English refers to size – to discuss brightness is all very confusing. Blame the Greek astronomer Ptolemy. His greatest contribution to astronomy was the multi-volume *Almagest* (The Great Book). As well as loads of exciting stuff about planets, the Moon, and the Sun, Ptolemy's work also contains a catalogue of around 1,000 stars visible to the naked eye (telescopes were yet to be invented).

Ptolemy divided stars into six groups, or magnitudes, with magnitude 1 stars being the brightest and magnitude 6 stars being the faintest. Each magnitude was considered to be twice as bright as the next one.

Working in England in the 1700s, Sir William Herschel – the astronomer who discovered Uranus (no jokes please) – thought that the Ptolemaic system of magnitudes wasn't sophisticated enough to catalogue heavenly bodies and proposed a new system in which a star with a magnitude 5 more than another star was 100 times brighter than it. The scheme wasn't widely adopted.

Eventually, British astronomer Norman Pogson (1829–1891) suggested using the same basis as Herschel and using decimals as well as whole numbers to specify magnitude. Astronomers everywhere eventually adopted this system. Pogson's system means that a difference of 1 magnitude is equal to a difference in brightness of the fifth root of 100, around 2.512. This number is now known as *Pogson's ratio.* Pogson's scheme originally used the northern pole star, Polaris, as a reference, giving it a magnitude of 2. However, Polaris turned out to be a variable star (see the section 'Mixing it up with variable stars') and now scientists use a standard measurement of brightness instead.

By arbitrary definition, the absolute magnitude is defined as the brightness of the object as observed from a distance of 10 parsecs with no interstellar dust or other intervening stuff that may dim the light from the star.

Knowing a star's absolute and apparent magnitudes (or just its magnitudes for a specific wavelength of visible light) enables astronomers to work out how far away a star is – after they've used some nifty maths, of course.

Mixing it up with variable stars

If you look at the sky night after night, little appears to change in terms of the stars. The Moon goes through its phases and the other planets zoom around the sky, but the stars seem unchanging.

In fact, the stars are all moving – some at huge speeds relative to the Earth. However, because the stars are so far away, this movement isn't observable on the Earth to the naked eye during the course of 24 hours.

But some stars, known as *variable stars,* do change their brightness in noticeable ways according to regular cycles. For example, the constellation of Cepheus, named after the mythological Greek king, includes the star Delta Cephei, which cycles through a range of brightness every 5.37 days. During this period, Delta Cephei's brightness more than doubles before dimming again. In the case of Delta Cephei, the regular expansion and contraction of the star, driven by nuclear fusion of helium within the star's core, drives this variation. (Chapter 12 explains the process in detail.) Scientists refer to variable stars that vary for this reason as *Cepheid variables.*

Not all variable stars change their brightness by expanding and contracting their outer layers like Delta Cephei. Beta Lyrae (the second brightest star in the constellation of the harp, or *lyra* in Latin) is known as an *eclipsing binary.* This star is called a binary because Beta Lyrae is in fact two stars that orbit each other on a regular basis. These binary star systems are very common but often the stars are so far away that you can only make out a single point of light. When one star passes in front of the other in its regular orbit – and that orbit has to lie in the direction of the Earth for us to be able to notice the variability – the star nearest the Earth eclipses part of the surface of the star behind. If the two stars have different brightnesses then this is observed from the Earth as a variation in the overall brightness of the system.

Measuring Stellar Distances

Look out of your window and pick an object, such as a lamppost at the end of your garden. What is the distance to that object? Although you can make a good guess (say, 10 metres), you can make certain of the distance in several ways.

- If you want, you can go outside with a tape measure and stretch it from your window to the lamppost.
- If you're really bored, you can use one of those laser rangefinders.
- If you're bored and rich, you can use a radar.

So how do you measure the distance to the stars? Clearly, using a tape measure is out of the question, and the other two methods are problematic as well. Even if you were able to guarantee that a laser beam would reflect back from the star's surface (which you can't), the beam would take years to send out and then travel back to the Earth, because lasers are limited by the speed of light. And a radar signal is too weak to get anywhere near the Sun, let alone another star.

Thanks, Eta Aquilae

Delta Cephei was not the first Cepheid variable that scientists discovered. That honour goes to the star Eta Aquilae, the fifth brightest star in the constellation of Aquila (the eagle). The brightness of Eta Aquila changes over a regular cycle of 7.2 days, and the British astronomer Edward Pigott noticed its variability in 1784. Pigott's friend John Goodricke discovered the variable nature of Delta Cephei, after which this type of variable star is now named.

Engaging in parallax thinking

One of the earliest known methods for measuring the distance to faraway objects – such as the distance from the Earth to a star – is to measure the star's *parallax,* which is an angular measurement usually represented by the Greek letter π (pi).

- You can see what parallax means by holding a finger up in front of your nose and closing first one eye and then the other. Your finger appears to move relative to the other things more distant in your line of sight.
- You can also see the effects of parallax when you're in a moving vehicle and look at objects that you're passing. Closer objects appear to move much faster than more distant ones.

Figure 5-4 shows another example of parallax in action. Say that you want to measure the distance to some far-off object – the lamppost at the end of your garden. Some distance beyond the lamppost you can see a row of trees. Figure 5-4 shows your pair of eyes near the bottom, while the lines indicate the views you get when you close one eye and then the other.

Consider the triangle made up by your eyes and the lamppost for a moment. Drawing a straight line from your nose through the lamppost to the trees cuts this triangle equally into two smaller triangles (always presuming, of course, that you haven't got oddly spaced eyes). These smaller triangles have a right angle at one corner, which enables you to use school trigonometry to calculate the angles, including the angle θ in Figure 5-4 (which happens to be half of the parallax π).

So if you know the distance between your eyes (measurable) and the parallax angle (also measurable), you can work out a pretty good approximation of the distance to the lamppost.

In case you're wondering, you can use an instrument known as a *quadrant* to measure an angle in situations such as the one that Figure 5-4 depicts. Surveyors use quadrants all the time to measure distances, hence this method is also known as the *surveyor's method.*

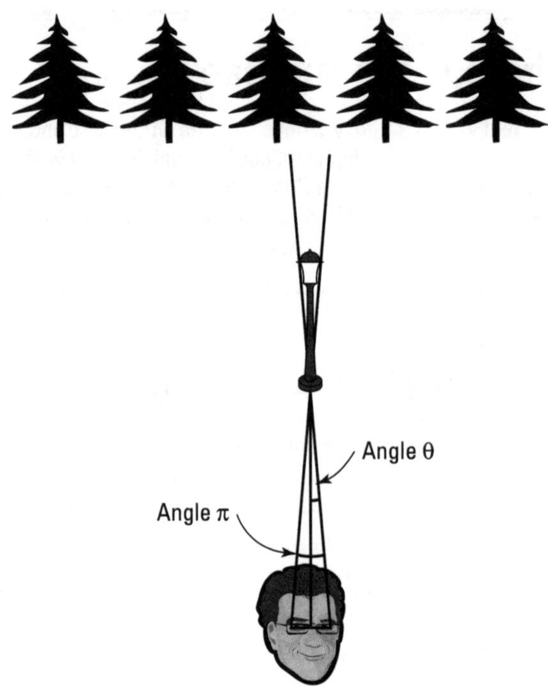

Figure 5-4: Simple diagram of parallax.

Applying parallax to stars

Sadly, because stars are so far away, the distance between your eyes isn't enough to yield a measurable parallax angle using the technique described in the preceding section.

However, because not all stars are the same distance from the Earth, you can measure the parallax of a star relative to two stars that are farther away. As Figure 5-4 shows, the two points in measuring the parallax don't have to be eyes; they can be any two observation points that are a known distance apart. In fact, the longer the distance between the two observation points, the more likely you are to be able to measure the parallax of very distant objects, such as stars.

What about making observations from opposite sides of the Earth? That may seem like a good baseline, but the distance just isn't great enough. Similarly, you can consider making one observation here on the Earth and another from a telescope based in space. This idea is good as well, but in fact you don't need to leave the Earth to identify two observation points that are appropriately far apart.

Instead, scientists can use the annual orbit of the Earth around the Sun to establish the two observation points. Figure 5-5 shows the Earth orbiting the

Sun and a star whose distance you want to measure. Because the Earth takes a year to orbit the Sun, consider two observation points six months apart that are at opposite points of the orbit. If you measure the position of the star in the night sky compared to some distant stars, the star's movement traces out a path. By measuring the parallax between two observations spaced six months apart, you can use the parallax formula to calculate the distance to the star.

Using parallax has problems. Figure 5-5 isn't to scale and because the distance for even the closest stars is huge compared with the distance from the Earth to the Sun, the parallax angle is incredibly small and thus hard to measure, so we can only use it for the very closest stars to us.

We can also make the maths a bit simpler. The average distance between the Earth and Sun, 149.6 million kilometres (93.5 million miles), is often referred to as an astronomical unit or AU. Using this as the basis of our measurements, then the distance to a star is equal to the reciprocal of the parallax. If the parallax is measured in the convenient unit of the arc second – 60 arc minutes are in a degree and 60 arc seconds in an arc minute – this gives the distance to the star in what astronomers call *parsecs*.

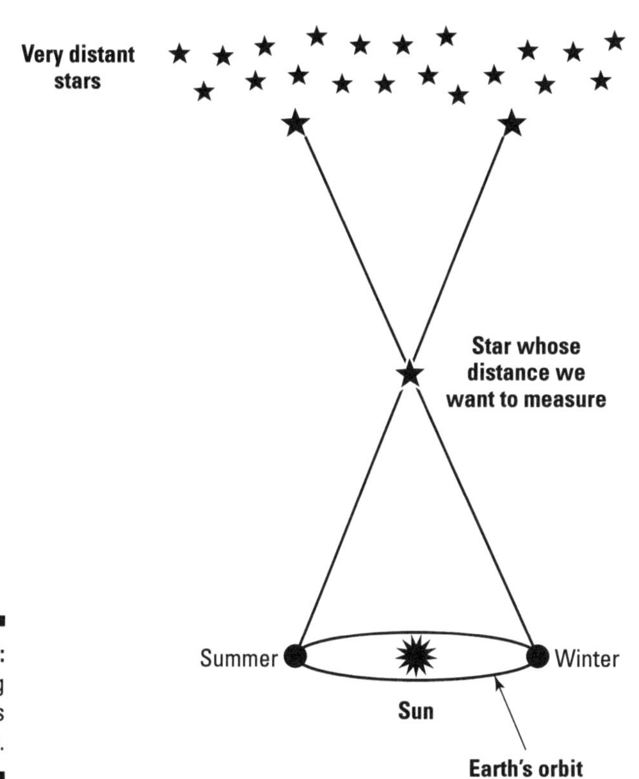

Figure 5-5: Measuring a star's parallax.

One parsec is the distance at which a star's parallax is one arc second and is around 3.26 light years, the distance light travels in 3.26 years or 31 trillion kilometres (19.3 trillion miles).

The best parallax measurements come from the Hipparcos space mission, which ran from 1989 to 1993 and measured the parallax angles of 120,000 stars very accurately and more than a million others with lesser accuracy. Considering that the universe probably contains more than 70 sextillion (70 thousand million million million) stars, scientists have some way still to go.

Measuring distance with standard candles

Thanks to the back-breaking research of a diligent woman, scientists have another means of measuring stellar distances for those stars beyond the ones that are close enough to have observable parallax.

Henrietta Swan Leavitt, an American astronomer who worked at the beginning of the 20th century, had the unenviable task of recording the magnitudes of thousands of stars on photographic plates at the Harvard Observatory. Her work focused on the clumps of stars known as the Small Magellanic Cloud, highly visible in the southern hemisphere.

Leavitt worked at a time when astronomy and physics were almost exclusively male domains. She spotted thousands of variable stars, many of them variable Cepheids (see the earlier section 'Mixing it up with variable stars'). During the process of cataloguing the stars, she realised that a simple relationship existed between the length of a star's variability cycle and its intrinsic *luminosity* (the rate at which a star emits energy).

Because all the stars in the Small Magellanic Cloud are roughly the same distance from Earth, Leavitt knew that the differences in their observed brightness are due to real differences in their intrinsic brightness. Knowing this, she realised that the more luminous a Cepheid variable star, the longer its variability cycle. Suddenly, armed with this information, scientists had another way to measure stellar distances: If you observe a star that you know is a Cepheid variable and measure the length of its variability cycle, you can work out its luminosity. From the luminosity you can then calculate a star's absolute magnitude and with observations of a star's apparent magnitude, work out how far away it is.

Following Leavitt's discovery, Ejnar Hertzsprung – famous for his star diagrams – measured the parallax of several nearby Cepheid variables, which enabled several distance calculations to be fine-tuned. As a result, Cepheid variables have become known as astronomy's so-called *standard candles* – objects that burn with a known brightness which can be used to compare with another object whose distance is unknown.

Shifting towards the red with the Doppler effect

Until someone tells you about the *Doppler effect,* the sound of an ambulance whizzing past can be very puzzling. Just at the moment the vehicle passes you, the siren changes from one sound to another. And this effect happens every time – spooky.

In fact, the sound of the siren is just an everyday example of the Doppler effect at work, a phenomenon named after a 19th-century Austrian maths whiz called Christian Doppler. The following sections explain how it works, and why cosmologists care about it.

The Doppler effect and sound

Sound is transmitted through the air as waves. These sound waves have peaks and troughs just like ripples on the surface of a pond when you throw in a stone, with the distance between successive peaks known as the *wavelength.* The sound waves that radiate out when you strike middle C on a piano have a wavelength of around 1.3 metres, for example.

In air, sound travels at around 330 metres per second. But when something that is making a sound is moving relative to the listener, you must also take into account the speed and direction of the movement. Therefore, because of the Doppler effect:

- When an ambulance is coming towards you, you hear sounds of a shorter wavelength and a higher pitch.
- When an ambulance is moving away from you, you hear sounds of a longer wavelength and a lower pitch.

The Doppler effect and electromagnetic radiation

Although the Doppler effect doesn't seem at first to have much to do with the universe, you can experience the phenomenon for all types of wave, not just sound waves. Electromagnetic radiation, including light, travels along as a wave, and you can observe the Doppler effect here as well.

But a slight complication arises when you consider how the Doppler effect affects the wavelength of electromagnetic radiation, and this snag is because of relativity (refer to Chapter 4 for more on relativity). As soon as you consider things travelling close to or at the speed of light, you can't simply add the velocities together as you do with the moving ambulance and moving sound waves.

Why is the night sky mostly dark?

For centuries, astronomers from Johannes Kepler to Sir Edmond Halley wondered why the night sky is largely dark. If the universe is infinitely large and contains an infinite number of stars distributed evenly, whichever direction you look, your line of sight should eventually end on the surface of star. This is often called Olbers' paradox, in memory of German astronomer Heinrich Olbers, who described it in the 1820s.

Because the night sky isn't completely bright, astronomers have argued that the universe is not infinite and doesn't contain an infinite number of stars. In fact, another effect is at work here. As Edwin Hubble showed in 1929, the light from all distant galaxies is red-shifted because the universe is expanding.

Even if the universe is infinite, Hubble's law says that the light from distant stars is red-shifted. Furthermore, the more distant a star, the more red-shifted the light becomes. At some distant point, any visible light emitted by a star red-shifts so much that you observe it on the Earth as invisible microwaves or radio waves.

To understand the Doppler effect on electromagnetic radiation, you need to do a little mathematics involving:

- The wavelength of the radiation being emitted
- The velocity of the object relative to yourself
- The speed of light (of course – it wouldn't be relativity without the speed of light in there somewhere!)

We won't spell out the formula here, but here's an example. If the thing you're looking at is moving away from you at 90 per cent of the speed of light, you find that the wavelength you see is more than four times the wavelength actually emitted.

Around the turn of last century, astronomers including Edwin Hubble (1889–1953) observed the light coming from distant galaxies (they were too far away for stars to be observed individually) and noticed that it was being *red-shifted,* which means the light was redder than expected. Electromagnetic waves act like all waves; when their source moves away from the receiver, the frequency with which they're received drops. For visible light, this effect shifts all light toward the lower end of the visible spectrum, toward red. The opposite effect, called *blue shift,* happens when a source of electromagnetic waves moves away from us.

Contemplating an Ever-Expanding Universe

As astronomers began to use clever methods to calculate distances to the stars, they realised that the universe was a lot bigger than previously thought. Soon they were to discover just how big it is – particularly based on the work of Edwin Hubble.

Probing the mysteries of nebulae

Scan the night sky with binoculars for a while and you may see fuzzy blobs mixed in among the stars and planets.

One well-known blob is M31 (the 31st such object catalogued by French astronomer Charles Messier), near the constellation Andromeda (named after a princess from Greek mythology), between the better known constellations of Pegasus the horse with its bright central square of stars, and W-shaped Cassiopeia.

M31 is obviously not a star or a planet, so what exactly is it? Its fuzzy appearance led astronomers to describe M31 as a *nebula,* a word that comes from the Latin for mist. Nebulae have been known for centuries; Persian astronomers observed them in the 10th century and Galileo Galilei took a good look at them with his telescope in the 17th century.

The exact nature of M31 and other such objects has been the subject of considerable speculation – the most prevalent idea being that nebulae were the birth place of stars or solar systems. To find out how their true nature was discovered, we turn our attention back to the stars.

Messier's heavenly catalogue

Frenchman Charles Messier first produced a definitive catalogue of nebulae in 1774. In fact, Messier was more interested in comets than nebulae and spotted 19 comets during his lifetime. His rationale for charting nebulae was so that he didn't confuse them with possible new comets, which also appear initially as fuzzy blobs similar to nebulae before they develop their telltale tails.

Messier's catalogue contains more than a hundred nebulae, which he designated with the letter M (for Messier) and a number. Many of the objects he catalogued are still known by their Messier designations today.

Parsons' drawings

The first signs that nebulae aren't just featureless fuzzy blobs came when William Parsons, the third earl of Rosse, made drawings of the nebula M51 in 1845. Parsons' drawings were made possible using the Leviathan of Parsonstown, a huge 17-metre long reflecting telescope with a 182-centimetre mirror, which he built at his family home of Birr Castle in Parsonstown, Ireland. At the time, the telescope was the largest in the world.

Parsons' drawings of M51 clearly showed that it had a complicated spiral structure and contained stars. In a paper that he presented at the time, he wrote: 'In the exterior stars of some clusters there appears to be a tendency to an arrangement in curved branches, which cannot well be unreal, or accidental.'

The shifting nebulae: Vesto Slipher

Vesto M. Slipher's biggest contribution to astronomy – and the fledgling science of cosmology – came in 1912 with his observations of nebulae and his research centred on measuring the blue-shift of M31.

Slipher, born in Mulberry, Indiana, in 1875, was the first person to graduate from the newly founded department of mechanics and astronomy at Indiana University in 1901. He was awarded the first doctorate from the department eight years later.

For his work, Slipher exposed several photographic plates at the Lowell Observatory in Flagstaff, Arizona, over long periods of time (because the nebula is so faint) to obtain a spectrum. By comparing the absorption lines in this spectrum with some standard reference spectra, he worked out how much the lines had shifted. The answer was surprising – M31 was moving towards the Earth at a velocity of 300 kilometres per second (187 miles per second), the highest speed ever recorded for a nebula.

Does this mean what you think it means? Yes, one day, M31 may collide with our own galaxy. But not just yet, probably in a few billion years or so. What might happen if it does is described in Chapter 13.

Amazing observations at Mount Wilson: Harlow Shapley

In the first three decades of the 20th century, California was the place to be if you were an astronomer – Pasadena in particular. The world's most powerful telescopes at the time were perched atop the 1,742-metre (5,715-foot) high Mount Wilson. Nebulae were an obvious target for Mount Wilson's incredible instruments, and they provided some exciting results.

Beginning in 1908, observations showed that the Andromeda nebula M31 has a spectrum that closely matches the Sun's, while other nebulae showed hints that they aren't just featureless blobs but rather concentrations of matter. The conclusion that these nebulae contain stars soon followed.

Harlow Shapley joined the Mount Wilson staff in 1914 and used the facility's instruments to look at globular clusters and work out how far away they are. (Globular clusters had been known since the mid-17th century; Edmond Halley, most famous for his work on comets, spotted one of the first – Omega Centauri – in 1677.) *Globular clusters* are collections of thousands or even millions of stars that are held together into a vast ball, which may be a few hundred light years in diameter, by their gravitational attraction.

Shapley paid particular attention to a globular cluster distributed in a symmetric halo around a point some 49,000 light years away in the direction of the constellation of Sagittarius. Shapley speculated that this point was the centre of our galaxy, the Milky Way, which extended roughly 326,000 light years in diameter.

Shapley went even further with his ideas. In the 1918 scientific paper 'Globular Clusters and the Structure of the Galactic System', he wrote that the 'globular clusters outline the extent and arrangement of the total galactic organisation'. Not only was he saying that the Milky Way is larger than anyone expected, but also that the globular clusters mark the boundary of not just the Milky Way but the entire universe.

He also argued that spiral nebulae – blurry, spiral-shaped objects that astronomers had observed – were unlikely to be separate galaxies of stars. In this, however, he was to be proven very wrong.

Dissenting with Shapley: Edwin Hubble

Astronomers have since come to accept much of what Shapley had to say about the shape of the Milky Way and the fact that the centre lies in the direction of Sagittarius. But not everyone was convinced by Shapley's conclusions about the spiral nebulae.

Among them was Edwin Hubble. Missouri-born Hubble studied mathematics and astronomy as an undergraduate but went on to become a Rhodes scholar in law. On his return from Oxford, he was called to the bar. But after the First World War, he realised that his true calling was astronomy and took a job at Mount Wilson in 1919.

One of Hubble's earliest pieces of work at the observatory was to draw up a classification sequence for nebulae. William Herschel and Max Wolf had previously drawn up their own, but Hubble believed their classification schemes to be inadequate. Herschel's scheme labelled nebulae from bright to faint while Wolf's identified 23 shapes, labelling each with a letter.

Expanding Hubble's vision of the universe

For most astronomers, coming up with one flash of inspiration during their professional lives would be enough, but not Hubble. His most far-reaching discoveries were still to come.

Hubble's Law

In 1929, Hubble published a landmark paper, 'A Relation between Distance and Radial Velocity among Extra-Galactic Nebulae', which built on the measurements of velocity of nebulae made by Vesto Slipher and others.

Hubble's first leap of inspiration was to use Cepheid variable stars (see the earlier section 'Mixing it up with variable stars') as standard candles to work out the distance to several spiral nebulae which yielded distances far beyond the known diameter of the Milky Way. This discovery demonstrated beyond doubt that many of these spiral nebulae lay beyond our galaxy, and paved the way for the possibility of galaxies other than our own. Hubble changed forever our view of the universe.

As Hubble measured more galaxies in this way, he realised that nearly all of them were moving away from us, a finding totally at odds with the idea of a static, uniform universe. In fact, the farther away the galaxies were, the faster they moved away from us.

As he put it in his paper: 'The results establish a roughly linear relation between velocities and distances.' Hubble called the constant slope of this line K and calculated its value as 500 kilometres (310 miles) per second per million parsecs (a figure that later proved to be wrong).

Putting it another way, you can say there is a direct linear relationship between the velocity with which a galaxy is moving away from us, and its distance from us. A galaxy twice as far away moves away from us twice as fast; if it's 100 times farther away, it moves away 100 times as quickly. This relationship is known as Hubble's Law.

The slope of the line that defines the size of the relationship between distance and speed is known as the *Hubble constant,* and it is a central number in cosmology. Why? It sets the rate at which the universe is expanding.

Seeing red, not blue

Hold your horses for a moment, you may be saying.

Slipher's observations (see the earlier section 'The shifting nebulae: Vesto Slipher') suggest that the Andromeda galaxy is approaching the Earth very quickly – and as such it should have a blue-shift, not a red-shift. Slipher's observations don't at first glance seem to fit with Hubble's Law.

In fact, astronomers have now discovered several objects that have blue-shifts and are approaching the Earth, rather than zooming away from it at incredible speeds. What sets these objects apart from the red-shifted objects is that they are all very close (in cosmological terms) to the Earth. (Astronomers have studied the Andromeda galaxy so much because it's the closest galaxy to the Milky Way and is visible to the naked eye.)

Gravity is responsible for the apparent contradiction. Newton's work states that gravity exists everywhere and that objects with mass attract each other. The Sun and the Earth are attracting each other in their elliptical dance, but the power and effects of gravity don't stop there. Galaxies attract each other as well. This local gravitational attraction means that clusters of nearby galaxies tend to stick together and interact, resisting the expansion of the universe on larger scales.

Think of the universe as a loaf of unbaked fruit bread you've left to rise. Galaxy clusters are like the raisins in the bread, gradually separating from one another as the cosmic dough expands.

A constantly expanding universe

So what does the Hubble constant tell us about the origins of the universe? Well, for a start, if galaxy clusters are currently moving away from each other, in the past they must have been closer together. Running the film backwards, you reach a point where everything is packed together very cosily indeed.

The Hubble constant can also tell us something about how long the expansion has been going on. With some clever computations using Hubble's original data, you end up with a period of time equal to about 2 billion years.

That tells us that at some point in the past – 2 billion years according to Hubble's original calculations – all these galaxies started off in the same place and until something crucial happened to give the object a huge velocity.

Since then, of course, scientists have gathered better data that suggest the Hubble constant is more like 50 or 100 kilometres per second per million parsecs, rather than the 500 Hubble himself calculated. This tallies with the roughly 13.7 billion years that cosmologists now think have passed since everything was packed together.

Still, the ramifications of Hubble's work were momentous. Scientists started wondering what got the expansion going. Perhaps this 'something' was some sort of explosion – or Big Bang even? Check out Chapter 6 for all about the Big Bang.

Chapter 6

Cooking Up a Big Bang

In This Chapter

- Envisioning the very small beginnings of an expanding universe
- Discovering ancient radiation from the early universe
- Naming and explaining the Big Bang
- Understanding where galaxies came from

If you ask the world's cosmologists how the universe began, most tell you that it started as a tiny, dense, incredibly hot fireball which expanded quickly outward to form the universe you know and love.

This explanation, known as the *Big Bang theory* (sometimes called the Big Bang model), stands head and shoulders above all other suggestions simply because it's supported by a wealth of calculations, observations, and – most crucially – relic waves of electromagnetic radiation from billions of years ago, known as the *cosmic microwave background* (CMB). (If that sounds like a special feature on a new-fangled kitchen appliance, read on.)

In this chapter we explore some of the evidence supporting the Big Bang. Along the way we meet some remarkable 20th-century scientists who helped to paint this compelling picture of the origins of time and space – including old friends Albert Einstein and Edwin Hubble (see Chapters 4 and 5, respectively) plus a host of others.

Gathering the Ingredients for an Expanding Universe

From time to time, you may wonder

- How big is the universe?
- What shape is the universe?
- How and when did the universe begin?

Perhaps you ponder these questions while strolling under the stars at night. Or maybe you struggle to find answers when your children pose them at bedtime.

Well, rest assured. These questions have also puzzled scientists for millennia. In fact, they are some of the central questions in the science of cosmology, and answers only began to emerge in the 1920s.

The answers came in two forms:

- Theoretical predictions based on Einstein's general theory of relativity
- Real-world observations that began with Hubble's measurements of the movements of galaxies

Predicting an expanding universe

Until the early decades of the 20th century, most scientists believed that the universe was a fixed and unmoving place, the permanent background to the wanderings of planets and stars.

But in 1916, when Einstein published his general theory of relativity (which we describe in more detail in Chapter 4), the idea of an expanding universe raised its pretty head.

When Einstein began to work through the implications of the equations behind the general theory of relativity, he came to a worrying conclusion. The calculations showed that the universe was *dynamic,* which means that it preferred to either expand or contract. Whatever it did, the universe as described by Einstein's equations was not static.

Einstein's biggest blunder

The model that Einstein formulated accommodated a universe being pulled together by gravity or a universe being thrown apart by some explosion in the distant past. But he didn't like the results that his equations were giving him. An expanding or collapsing universe didn't tally with what astronomers saw in the skies above them – that the universe was pretty much the same wherever you looked, seemed to have been around forever and, other than the occasional birth or death of a star or galaxy, appeared to be unchanging.

So Einstein decided that he needed to tweak the model so that it matched the perceived reality. To do this, he added a term to his equations that represented a force of repulsion to counterbalance the pull of gravity. He called this term the *cosmological constant.*

Later, though, after the evidence showing that the universe really was expanding had convinced Einstein, he allegedly referred to the cosmological constant as his 'greatest blunder'.

Observing the expanding universe

Other scientists soon realised the implications of Einstein's general theory of relativity for the expansion of the universe, even if the great man himself wasn't willing to consider them (see the sidebar 'Einstein's greatest blunder'). Notable among these scientists was the mathematician Alexander Friedmann.

Alexander Friedmann was a Russian scientist who lived most of his life in St Petersburg (or Leningrad as it eventually became). Friedmann's big contribution to cosmology was finding solutions to Einstein's equations.

What we mean by finding a solution here can be illustrated by the equation $x^2 - 6x + 8 = 0$. There are two, and only two, values of x that work in this equation: 2 and 4; try it yourself. Any other values – try 3 for example – give the wrong answer.

What Friedmann found was that when he assumed that the universe was homogeneous and isotropic, the solutions to Einstein's general relativity equations all showed that the universe was expanding.

But the calculations of Friedman and others weren't really taken seriously until 1929, when astronomer Edwin Hubble and his colleagues made their fantastic discoveries about the existence of other galaxies beyond the Milky Way and the movement of those galaxies *away* from one another (see Chapter 5).

The more galaxies that Hubble measured and observed, the more apparent it became that most galaxies are moving away from the Earth, no matter which direction you look. In fact, the farther away the galaxies are, the faster they seem to be moving away.

Showing that the universe is expanding was an enormous breakthrough. It was so big that some have called this discovery the greatest of 20th-century science!

Understanding expansion, or inflating the universal balloon

Hubble showed that the universe is expanding, but what does that mean exactly?

Remember Woody Allen's film *Annie Hall?* In one scene, the young character Alvy is in the doctor's office worrying about the expansion of the universe. 'The universe is expanding,' he explains. 'The universe is everything, and if it's expanding, someday it will break apart and that will be the end of everything.'

Alvy's mother interjects: 'He's even stopped doing his homework . . . What has the universe got to do with it? You're here in Brooklyn. Brooklyn is not expanding!'

Thankfully, when scientists talk about the universe expanding, they're not talking about people, parts of New York, planets, galaxies, or even clusters of galaxies expanding. Chemical and gravitational forces that are stronger than the expansion hold all these objects together.

When scientists refer to the expansion of the universe, they're thinking about enormous scales – bigger than giant clusters of galaxies. These clusters of galaxies act as markers of expansion.

One simple way to think about the expansion of the universe is to imagine small stickers on the surface of an expanding balloon. The stickers are the clusters of galaxies and the balloon is the expanding universe. As you gradually blow up the balloon, all the stickers get farther apart, but they don't get any bigger themselves.

Another way to envisage an expanding universe is to think of an elastic bracelet strung with beads. As you stretch the bracelet wider and wider, the beads move away from one another. The elastic in this case stands for space, and the beads represent galaxy clusters moving away from each other.

Of course, these analogies are only illustrations. The universe isn't really anything like a balloon or an elastic bracelet. But thinking about the problem like this helps you to see that the expansion of the universe isn't so much about galaxies moving away from each other through space: It's about the space between galaxies expanding or swelling.

Turning Up the Heat on Expansion

Scientists in the 1930s realised that if the universe was expanding – as Einstein's formulas predicted and Hubble's measurements showed – at some point in the past the universe must've been smaller than it is now.

If you were to turn back the clock of cosmological time, you would eventually reach a point, 13.7 billion years ago, when all the matter in the universe was packed together more densely than it is today.

In the early days of modern cosmology, the most influential supporter of the idea that the universe began small was Georges Lemaître, a Belgian priest and graduate of the Massachusetts Institute of Technology. Many scientists now regard Lemaître as the father of Big Bang cosmology, because he proposed in 1927 that the universe had a beginning in the form of a *space particle* or *primeval atom*.

When Albert met Edwin

In 1931, Albert Einstein paid a visit to the Mount Wilson Observatory in California, which was home to some of the most powerful astronomical equipment in the world. During this visit, Einstein met Edwin Hubble for the first time.

As Hubble explained the results that he had obtained from studying the galaxies moving away from the Earth, Einstein admitted that the cosmological constant that he had introduced in a vain effort to show the universe was static (see the sidebar 'Einstein's greatest blunder') was a major scientific goof.

Einstein's wife Elsa also came on the trip, apparently in a less modest mood. When the scientists explained to her how their enormous 2.5-metre (100-inch) telescope was able to determine the structure of the universe, she was distinctly unimpressed. 'My husband does that on the back of an old envelope,' she told them!

Of course, an object that's infinitely dense and yet has a size of zero is very difficult to conceptualise – even for scientists! But the following section endeavours to tackle this most challenge of topics.

Starting off small

Points of infinite density such as the one that scientists believe served as the start of the universe are known as *singularities*. Scientists don't like singularities very much, which is understandable because singularities cause all the theories of science to break down.

So, when scientists make their calculations about what happened during the Big Bang, they generally begin their descriptions at a point just after the birth of the universe – at a phase when everything was contained in a minuscule, hot fireball.

How small was that fireball? Consider the full stop at the end of this sentence and how small it is in comparison to the size of the entire universe. The cosmic fireball is that factor smaller than the full stop and then some. We're talking truly minute!

Naming the Big Bang

At first, some physicists resisted the concept of a universe that started from nothing, because many of them didn't like the idea that time had a definite beginning. Those researchers proposed alternative models to the Big Bang, some of which we describe in Chapter 8.

One of the main opponents to the idea that the universe had a definite beginning was the British astronomer Fred Hoyle. He, along with Hermann Bondi and Thomas Gold, proposed instead a *steady state universe* in which tiny amounts of new energy were being created over time, to fill the gaps as other galaxies moved farther away from each other.

Ironically, Hoyle coined the term 'Big Bang' in a 1950 BBC radio interview as a flippant name for a theory he didn't believe in. Nevertheless, the insult caught on and the term is still used today, despite the fact that the beginning of the universe wasn't big – in fact it was vanishingly small – and no bang or explosion was involved.

Checking the Oven: Looking for Fossil Radiation

In the 1940s, the most important advocates of the Big Bang theory were George Gamow and his colleagues Ralph Alpher and Robert Herman. They realised that if the universe had begun in a hot, dense state then some radiation should be left over in the universe recording this explosive beginning. This is because much of the radiation from the Big Bang hasn't had a chance to interact with anything else in the emptiness of space and therefore remains unchanged (other than cooling due to the expansion of the universe).

Alpher and Herman predicted that the *fossil radiation* of the Big Bang – radiation that cooled as the universe expanded – should now have a temperature of around 5 degrees above absolute zero (absolute zero is the same as –273 degrees Celsius), otherwise known as 5 kelvin (K).

Sensing the radiation

The first proof that this remnant radiation exists came in 1964. That year Arno Penzias and Robert Wilson, two radio engineers from the Bell Telephone Laboratories in New Jersey, were working to fine-tune a horn-shaped radio antenna designed to detect microwaves.

Like light, *microwaves* are a kind of electromagnetic radiation. Microwaves have a frequency of around 10,000 million waves per second, whereas visual light has a frequency ranging from 450 to 750 trillion waves per second. This is the same type of radiation used in microwave ovens. The microwaves increase the energy of the particles in the food, thereby raising their temperature.

Penzias and Wilson were investigating microwaves because they wanted to open up a new form of communication; they needed to figure out how much background noise microwaves would generate. Although the two scientists diligently accounted for all possible sources of background radiation, they soon became concerned because their radio antenna was picking up too much microwave radiation. At first they thought that some problem with their equipment was the cause. In desperation they even considered whether pigeon droppings in the antenna were perhaps causing the problem. (Sadly, cleaning out the instrument made no difference.)

Eventually, they realised that the extra radiation was the same no matter where in the sky they pointed their antenna, and so they knew that the background noise must be coming from outside the atmosphere.

The researchers also found that the excess microwaves were the same during the day and night, as well as at different times of the year. Because the Earth was rotating on its own axis and also around the Sun, the microwaves must be coming equally from all directions in the sky. This suggested that the microwaves must come from beyond our solar system – in fact, from beyond our galaxy.

Around the same time, scientists at nearby Princeton University were also beginning to look for fossil radiation from the Big Bang, having recently recalculated the figures that Alpher and Herman had published years before.

When the Princeton team got wind of the radiation that the Bell Labs team were picking up, they realised they'd been scooped. The Princeton team knew exactly what the excess microwaves were – evidence of the Big Bang.

The microwaves that fill the universe at a temperature of 2.73 kelvin, first discovered by Penzias and Wilson, are the most important piece of observational evidence supporting the idea of the Big Bang. Scientists call them the *cosmic microwave background,* or CMB. The CMB is the afterglow of the Big Bang, coming to the Earth from a time when the universe was a thousand times smaller than it is today, long before planets, stars, or galaxies existed.

Putting a time to the CMB

Scientists calculate that the CMB originates from roughly 380,000 years after the Big Bang. At that time, the temperature of the universe was about 3,000 kelvin, which is hot enough to send out radiation in the ultraviolet spectrum.

So why do scientists pick up microwaves nowadays? The expansion of the universe since the Big Bang has caused that original radiation to be *red-shifted* (see Chapter 5) all the way through the electromagnetic spectrum down to cool microwaves.

During the first 380,000 years following the Big Bang, the universe was too hot for *photons,* the elemental particles that carry electromagnetic energy, to move around unimpeded. The universe was even too hot for atoms to exist because electrons would have been stripped away from the nucleus by the heat. Chapters 9 and 12 cover these basic building blocks of all material in greater detail.

In fact, if you were able to look across the universe during its first 380,000 years or so, all you would see was a bright glowing fog in all directions. Any photons hitting your retina would have bounced off nearby electrons just fractions of a second before you observed them, a process known as *Thomson scattering* (named after the physicist J.J. Thomson who first explained it).

Only after the universe cooled down enough for protons to permanently capture electrons, thereby forming neutral hydrogen, did the fog lift and the cosmic background radiation become released.

The point when this happened is sometimes called *the time of last scattering* because it was the last time most of the CMB photons directly scattered off matter.

Reading the CMB

The cosmic background radiation that scientists detected in the 1960s seemed to be uniform wherever they pointed their instruments. For this reason, cosmologists say that the expansion of the universe is *isotropic,* meaning that it's the same in every direction.

A little contemplation reveals two possible explanations for this situation:

- If everything's moving away from us, perhaps the Earth is at the centre of the universe.
- Perhaps the universe looks the same when viewed from any other galaxy as well.

Scientists prefer the second answer, partly out of modesty – the fact that Earth happens to be at the centre of everything seems unlikely.

In fact, most cosmological models are based on a formal statement called the *cosmological principle.* This fundamental rule says that the universe, at least on large scales, looks the same in all directions (hence it's isotropic) and has the same properties in every place (it's *homogeneous*).

Finding variation in the CMB

Of course, a casual glance up at the night sky reveals that the universe isn't isotropic in detail. Stars appear in clumps with galaxies sprinkled here and there. On scales of less than about 300 million light years, the universe is really pretty clumpy – neither homogeneous nor isotropic. But as astronomers look farther and farther afield in the universe, there appears to be no overall pattern to the distribution of galaxies – it's essentially smooth, with very small deviations from that smoothness.

Whenever cosmologists see something that reveals the universe isn't exactly the same in all directions, they call it *anisotropy.* One example of anisotropy is the distribution of stars in the sky, and another is in the CMB.

Although the CMB is to a large extent the same no matter which direction you measure it from, recent experiments show that there are tiny variations in its temperature.

Recent experiments such as those performed by NASA's Cosmic Background Explorer (COBE) and the Wilkinson Microwave Anisotropy Probe (WMAP), both of which we describe in Chapter 20, show that the smoothness of the microwave radiation actually contains subtle variations in temperature, at the level of 1 part in 100,000 kelvin.

One part in 100,000 is a tiny level of variation. In terms of smoothness, the universe is smoother than a snooker ball. In fact, if the Earth were as smooth as the temperature of the CMB, the whole world would be flatter than the Netherlands.

The sensitive equipment on the COBE satellite was able to pick up these small spots that were ever so slightly colder or hotter than average on its map of the CMB. When the COBE data was first made public in 1992, the scientific world was thrilled to see evidence of *anisotropies* – or wrinkles in time as they're sometimes erroneously called (although they do point to events far back in the history of the universe). The physicist Stephen Hawking called the COBE data, 'The greatest discovery of the century, if not of all time.' Clearly, these patches of hot and cold were a big deal! But why? And what do they mean for understanding of the Big Bang?

Well, the quick explanation is that small variations in the CMB represent the ancient 'seeds' from which galaxies grew. They are evidence of the earliest clumping together of matter in the early life of the universe before it was 380,000 years old.

These minuscule variations in the amount of microwave photons from place to place in the universe have been most accurately measured by WMAP. Its map of the variations shows these variations as light hotter patches and darker colder patches (although the variation was very small – just 1 part in 100,000), as the photograph in the colour section shows.

Identifying the source of CMB variation

So where did these wrinkles or variations in the CMB come from?

Scientists think that the first beginnings of structures in the universe started as tiny fluctuations in matter at the quantum level. A *quantum fluctuation* is a temporary change in the amount of energy at a point in space. Think of quantum fluctuations as minuscule, momentary irregularities that pop into existence before disappearing again. These energy changes take place according to Heisenberg's uncertainty principle (see Chapter 9). The length of time the fluctuation exists is inversely proportional to the amount of energy involved.

What we see today when we observe the CMB is the pattern of fluctuations that were present at the moment when matter started to dominate the universe at the age of 380,000 years. This is because most of the photons since then haven't interacted again and so retain the signature of the structure of the universe at that point.

These fluctuations weren't unique to the early universe. In fact, according to Heisenberg's uncertainty principle, the universe isn't perfectly regular, even at the smallest scales. So scientists can safely say that quantum fluctuations are happening all around you right now, although they have no noticeable impact on human scales. They're simply too tiny.

But in the early universe, just a fraction of a second after the Big Bang, a process called *inflation* (which you can read about in Chapter 7) stretched and magnified the quantum fluctuations enormously before the fluctuations blinked out of existence.

Incidentally, quantum physics has some precise predictions about the way these fluctuations would have been distributed. In fact, one of the most convincing pieces of evidence in support of this idea is that the anisotropies in the CMB and the distribution of galaxy clusters in the universe today match pretty neatly.

Observing blackbody radiation

The CMB has another important characteristic: When scientists measured its spectrum (see Chapter 5) they found that it has the kind of energy distribution known as *blackbody radiation.*

If you measure the radiation coming from a bright object like a star, for example, and plot it on a graph, you find that the object emits radiation across a range of wavelengths. This measurement is called its *spectrum.* For most glowing objects (such as a star), the graph of their spectra is complicated, with a lot of different high and low points.

Black bodies aren't always black

The amount of radiation that black bodies emit is directly related to their temperature. So black bodies below roughly 430 degrees Celsius produce very little radiation at visible wavelengths and appear black to the human eye.

Above 430 degrees Celsius, black bodies begin to produce radiation at visible wavelengths, starting at red, continuing to orange, yellow, and then white before ending up at blue as their temperature increases.

At very high temperatures, black bodies again appear black to the human eye because the body emits most of its radiation as ultraviolet radiation and X-rays.

Understanding black body radiation is important for understanding the CMB and also stars, which exhibit spectra that are like black bodies but overlaid with spikes and troughs relating to the chemical elements that they contain.

But the spectrum of a special type of heat radiation, produced by objects that physicists call *black bodies*, has a simple, regular shape (shown in Chapter 9). A good example of a black body is an oven in which the inside has come to a precisely uniform temperature. Similarly, blackbody spectrum is only produced by systems in which temperature has reached equilibrium (that is, is no longer changing).

Fossil radiation is the best example of blackbody radiation known to humankind. Experiments performed by the Cosmic Background Explorer (COBE) satellite in 1990 confirmed that the CMB has exactly the spectrum of radiation that scientists would expect if it were produced in a cosmic furnace.

The implications of this discovery are pretty impressive. In fact, because the CMB has a blackbody spectrum, scientists can show that at one time the material of the early universe must have been spread evenly throughout space, at an even temperature and density.

In short, the discovery that the CMB radiation was blackbody radiation is a very convincing endorsement of the hot Big Bang theory, showing that the universe started in a small, hot, dense, and uniform state and then quickly expanded and cooled to reach its current state.

Appreciating the amount of energy involved

If the universe began hot and small, you may then wonder about the temperature involved. A good place to start, in fact the only place to start, is 10^{-43}

seconds after the Big Bang: That is, just 0.0000000000000000000000000000 000000000000001 of a second after things got started, when the universe was just a tiny part of a smidgen of a twinkling of a fraction of a second old.

Scientists call this moment the *Planck instant,* and it represents the point before which the laws of physics as we currently understand them, such as Einstein's general theory of relativity, cannot help us explain what was happening. At this time, the universe was roughly 10^{32} kelvin (100 million, trillion, trillion kelvin – also known as pretty darn hot!).

Things didn't stay that way for very long, however. The universe was quickly evolving, rapidly expanding, and cooling down, with the result that the fundamental forces of the universe began to disentangle themselves from one another, separating completely within just 10^{-12} seconds. Eventually, the stars and galaxies that fill the universe began to form.

Cosmologists are most interested in this very early period. Something momentous seems to have taken place. As well as the separation of the forces (which we discuss in Chapter 10), the universe experienced a brief but gigantic jolt, a sudden outward rush of expansion that took the quantum fluctuations in its structure and super-sized them.

Scientists call this momentary rush *inflation*, a topic that we explore in more detail in Chapter 7.

Chapter 7

Letting It Rise: Expanding and Inflating the Universe

In This Chapter

- Exploring the universe immediately after the Big Bang
- Figuring out the shape of the universe
- Explaining the universe's dramatic early expansion
- Relying on inflation to solve problems in understanding the universe

The Big Bang model (which we describe in Chapter 6) says that the universe began as a small, hot, dense place, and that ever since that moment it's been expanding.

Solid, real world evidence backs up the Big Bang as an explanation for the history of the cosmos. But in the second half of the 20th century, scientists realised that the model left a few questions unanswered. For example:

- **How did the temperature of the universe get to be the same everywhere?** The Earth is constantly bombarded with (harmless) microwave radiation from every direction with a temperature of 2.73 degrees above absolute zero.
- **Why is the universe so smooth and flat?** It may not seem so when you look through a telescope but the universe is actually incredibly uniformly distributed. In addition, the cosmos seems to have just the right amount of stuff for the universe to be finely balanced between expanding forever and contracting under the force of gravity.

In this chapter we discuss *inflation,* the remarkable theory that offers an elegant explanation for these questions and others. During inflation, for a split second almost immediately after the Big Bang, the universe expanded at breakneck speed – faster, in fact, than the speed of light. If you think that contradicts the laws of nature, read on to find out why it doesn't.

Going Back to the Beginning

The central idea of Big Bang cosmology is that the universe is expanding from a smaller, hotter, and denser state, which emerged roughly 13.7 billion years ago. For logical people, this theory implies that the universe started with everything packed into a single point of infinite density.

But problems emerge when scientists take things to extremes. These problems relate in part to the laws of *quantum physics,* the science of very small things. Quantum physics says that it makes no sense to talk about distances smaller than the so-called *Planck length,* which is about 10^{-35} metres long (that's 0.00000000000000000000000000000000001 metres). This is because Heisenberg's uncertainty principle (which we explain in Chapter 9) says that to measure such a minuscule distance would require the use of a very high energy photon. If this whizzy photon hit the object under scrutiny, scientists predict it would create a miniature black hole.

Thus any discussion of the expanding universe begins at the time when the universe was that small, when it had a staggering density of 10^{96} grams per cubic centimetre and a temperature of 10^{32} kelvin. Imagining something so dense and hot is nearly impossible. Lead, the very dense metal used to shield against radiation, only has a density of 11 grams per cubic centimetre and this is one with 93 noughts after it times more dense. The core of the Sun – a very hot place we think you'll agree – only has a temperature of 15 million kelvin. That's 1 with 26 noughts after it times chillier.

And when was the universe this size, density, and temperature? Well, to match the distance we mention just above, you must focus on a moment called *Planck time* – 10^{-43} seconds after the Big Bang.

At this great density, you may wonder how the universe began expanding at all. Wouldn't the combined gravity of all that mass drag the universe back together again? Well, cosmologists think that at this point something must have given the universe an enormous outward shove, expanding it dramatically in the blinking of an eye. This period, which lasted from about 10^{-37} to 10^{-35} seconds after the Big Bang and is known as *inflation*, has become a key part of our understanding of the very early universe.

Defining the universe

Before we move on to what caused inflation and how it explains a lot about the universe, we must take a moment to define the word *universe*.

- **The visible universe.** Perhaps the easiest way to define the universe is to say that it's made up of everything that you can see with all the various instruments available today – including all those high powered telescopes.

  At the beginning of the 20th century, the visible universe was limited to the Earth's own galaxy, the Milky Way, and a few close neighbours, cosmologically speaking. Now, the visible universe is much bigger thanks to improvements in technology.

  Defining the visible universe has some good points – after all, it limits discussions to the stuff that humans can actually see. However, focusing only on the visible universe also has some serious shortcomings.

- **The observable universe.** Another very important definition when talking about the universe is *the observable universe.* When scientists talk about the observable universe, they're referring to the space that it's theoretically possible to see according to the laws of physics – if humans were to build bigger and better telescopes.

Heading way back

In terms of time related to inflation, you need to start off by thinking about the time since the Big Bang: 13.7 billion years. In that time, by definition, light can have travelled a maximum of 13.7 billion light years.

You need to be careful here. If you were able to look out at the universe with an expensive telescope and see a galaxy that is, say, 12 billion light years away, you'd actually be seeing the galaxy as it was 12 billion years ago. Your view isn't where the galaxy is now. In the intervening time, the galaxy in question has become much farther away, courtesy of the expansion of the universe.

To put it another way, in the 12 billion years since light from this theoretical galaxy began its journey, the distance between the Earth and the galaxy has grown bigger. (Remember that when scientists talk about the expansion of the universe they mean that the space between the Earth and the distant galaxies is getting bigger. The galaxies aren't moving through space away from the Earth.)

In fact, scientists have calculated that the current distance between the Earth and the farthest observable thing is more than three times the speed of light multiplied by the age of the universe. That's roughly 47 billion light years. This amount is the current limit, or *horizon*, of the observable universe.

Pondering the Horizon

Of course, this situation doesn't necessarily mean that emptiness lies beyond the universe's horizon. The current horizon is defined only by humans' particular place in space. A galaxy 10 billion light years away from the Earth has its own horizon that partially overlaps the Earth's, but also includes parts of space that Earth-bound humans can't see.

Additionally, the size of the Earth's cosmological horizon is set partly by the fundamental law of physics, which says that nothing can travel through space faster than the speed of light.

When scientists started pondering the Big Bang, this limitation raised an important issue known as the *horizon problem,* which we discuss in this section. The horizon problem is the first of the problems that inflation helps solve.

Scientists realised that places existed within the Earth's observable universe that are separated by greater distances than light could have travelled since the Big Bang. This problem is easy to understand. If you look out into space in one direction, the oldest light that can reach you came from 13.7 billion light years away. When you look out into space in the opposite direction, the same applies.

But those two places in opposite parts of the sky are now 27.4 billion light years apart. There's no way that light – or anything else such as the exchange of heat to even out the temperature – can have travelled between these places since the Big Bang.

Think about this problem another way: Aliens living on a planet in an area that lies in one direction from the Earth would be unable to see aliens living in the region in the opposite direction, or vice versa – even though the Earth's own horizon that's in the middle includes them both.

Each of the circles in Figure 7-1 represents the cosmological horizon for the Earth, and for theoretical aliens living in galaxies A and B. Although we are close enough to A and B to see them both, they are too far separated in space for light to have travelled between them since the Big Bang. The fact that some regions in space exist that are too far apart for light to have travelled between them also means that nothing else can have travelled between them.

So what? you may ask. Well, when scientists measure the cool afterglow of the Big Bang – the cosmic microwave background that we describe in Chapter 6 – they find that all the observable universe is remarkably similar in temperature. If no time has existed for light or anything else to connect these various regions in the universe, how did they get to be so similar? This problem is what cosmologists call the horizon problem.

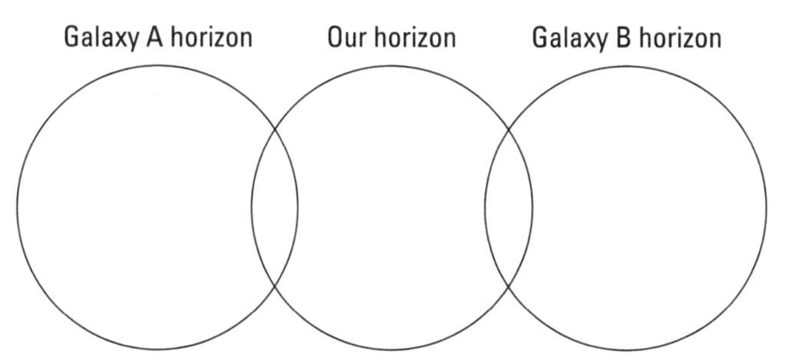

Figure 7-1: The cosmological horizon for Earth, and for theoretical aliens living in galaxies A and B.

Shaping the Universe

Another conundrum that scientists came across when they started measuring the physical properties of universe is called the *flatness problem*. This problem relates to something scientists have known since soon after Einstein introduced his model of the universe. They realised that the universe can have one of three overall shapes: closed, open, or flat (see Figure 7-2).

In 1922, Russian scientist Alexander Friedmann was the first to realise the universe's three possible shapes. Friedmann's work showed that if you treat the universe as *homogeneous* (the same at all places) and *isotropic* (looks the same in all directions), you can have only three possible outcomes for an expanding universe, depending on the average amount of mass within the universe (see Figure 7-3 for the three possible outcomes).

The following sections discuss these three possible shapes in greater detail.

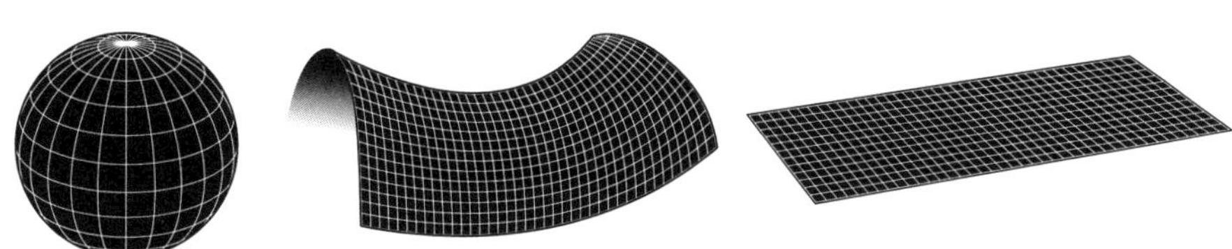

Figure 7-2: The three possible shapes of the universe.

Getting the squeeze

The first possible outcome for the fate of the universe – a *closed universe* – comes about when the overall density of matter in the universe is larger than a specific amount, which scientists call the *critical density*.

Cosmologists use the Greek letter omega (Ω) to refer to the ratio of the actual mass density of the universe to the critical density. If Ω is greater than 1, space is considered *closed*, and it will curve around on itself – a little bit like a three-dimensional version of the two-dimensional surface of a balloon, as shown in Figure 7-2.

For closed universes, the gravitational effect of all the matter will become enough eventually to act like a cosmic brake on expansion. The universe will gradually stop getting bigger, turn back on itself, and finish in a Big Crunch. Ouch!

Expanding forever

Another possible outcome of the universe – an *open universe* – comes about if the total density of matter in the universe is less than the critical density – that is, if Ω is lower than 1. In this case, the universe is considered *open* and it will become shaped something like a saddle, turning down in the middle and up at the ends.

In this scenario, the rate at which the universe is expanding will overcome the gravitational effect of all the stuff in it, meaning that the universe will keep expanding forever. In the end, this leaves the universe a cold and desolate place sometimes called the Big Chill (which we explain in Chapter 18).

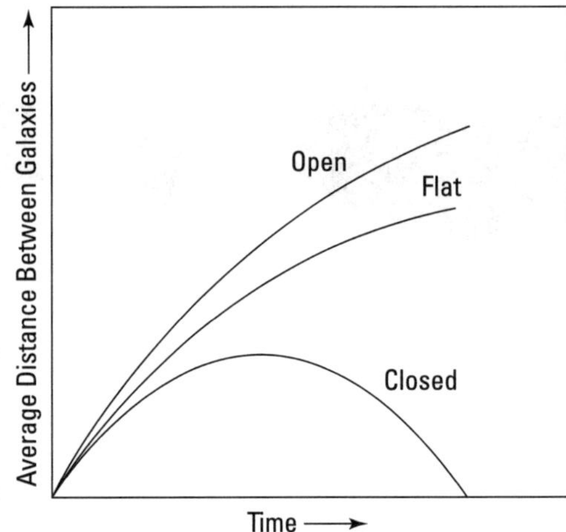

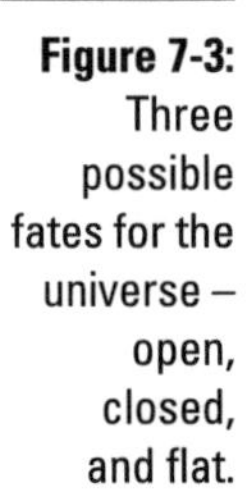

Figure 7-3: Three possible fates for the universe – open, closed, and flat.

Living in a universe that's just right

Balanced on the knife-edge between these two possibilities lies a third outcome, the so-called *flat universe,* a universe with no curvature at all and a density exactly equal to the critical density of matter.

A flat universe keeps expanding for some time until it reaches a kind of state of balance far in the future, where it then neither expands nor contracts.

To calculate the critical density needed for a universe to behave in this way, scientists need to take into account such things as the rate at which the universe is expanding (called the *Hubble value*), the speed of light, and other factors.

Based on all these factors, the best calculations so far give a value for the critical density of something being in the region of 10^{-29} grams per cubic centimetre, which corresponds to just a few hydrogen atoms per cubic metre, on average. This amount of matter is really tiny.

The remarkable thing is that as scientists have developed better ways to estimate the overall density of the universe, it has become clear that the universe is in fact tantalisingly close to being flat. The universe really does seem to have an average density that's close to the critical density.

This fact is pretty remarkable in itself – but even more so when you factor in a little more Einstein. You see, Einstein showed that any deviation from perfect flatness in an expanding universe would tend to get more pronounced as the universe expands. That is, closed universes tend to get more closed, and open universes get more open. So if only small deviations from the critical density exist today, these deviations must have been infinitesimally small at the time of the Big Bang.

The puzzle for cosmologists – the flatness problem itself – is how any small deviations could have stayed so small. Inflation is the best answer.

Imagining Inflation

At the beginning of the 1980s, a young American physicist called Alan Guth was pondering the horizon and flatness problems and realised that these problems disappeared if he assumed that the universe had suddenly expanded dramatically for a tiny fraction of a second just after the Big Bang. Guth called this idea inflation. In a scientific article he published in January of 1981, he explained that he didn't really have any solid evidence to support his theory, but that it 'seems like a natural and simple way to eliminate both the horizon and flatness problems'.

Expanding exponentially

The basic idea of inflation is straightforward. For a brief moment, when the universe was roughly 10^{-35} seconds old, the universe suddenly inflated by an enormous amount, a factor of 10^{30} or more, in about 10^{-32} of a second. The expansion for this flickering moment is thought to have been exponential, meaning it doubled in size roughly every 10^{-34} seconds.

That is, in less than the blinking of an eye, what is now the observable universe grew from being smaller than a tiny fraction of the size of a single atom to something bigger than the size of a grapefruit. We explain the mechanism Guth proposed for this in the later section 'Reversing gravity'. After reaching grapefruit size, the expansion (that observed by Edwin Hubble) continued as the universe cooled and the stars and galaxies we see around us were formed.

Scientists believe that the speed with which the universe expanded during inflation was faster than the speed of light. 'But,' you ask, 'surely nothing can travel faster than light?' Well, you're right if you limit yourself to things moving *through* space. In actuality, the laws of physics allow for space itself to expand more quickly than the speed of light. If this rate of expansion isn't against the law, the universe would have been mad not to try speeding that fast at least once, right?

Solving some tricky problems

In one fell swoop, Guth's inflation idea promised to solve the tricky horizon and flatness problems that plagued the Big Bang model.

- **The smoothness problem.** By dramatically expanding a tiny area of space into a much larger area, inflation could have smoothed out any existing curves, making the observable universe smooth. As well as producing extreme smoothness, inflation could also have left behind the kind of tiny ripples in the background radiation that grew to form the galaxy clusters seen today.
- **The flatness problem.** Inflation enables the universe's shape to be flat, rather than closed or open. To conceptualise how inflation does this, think of yourself standing on the surface of the Earth. As it is, the Earth seems pretty flat, but imagine if the Earth was blown up like a balloon to a stupendously bigger size. The surface would seem as flat as a plate to you – yet still have a overall three-dimensional shape to a distant observer – as should the observable universe according to inflation.
- **The horizon problem.** Accelerated expansion also gets around the horizon problem, by allowing the observable universe to have expanded from an area small enough for light to have travelled across it in the time since expansion began.

Reversing gravity

The inflation theory was neat, but other scientists wouldn't have taken it seriously if Guth hadn't offered a viable explanation for *how* it may have happened.

His suggestion was that the force of gravity acted in reverse for 10^{-32} of a second, pushing outward instead of pulling inward. He also came up with an explanation for how this *anti-gravity* may have come into being.

The basis of Guth's explanation is that early in the history of the universe, when energy levels were enormously high, matter took the form of *fields*.

No, not the kind of fields farmers drive their tractors over. When physicists refer to fields they're talking about assigning physical characteristics to points in space.

One example of a field is an electromagnetic field. Scientists think that in the early universe, matter was in the form of *scalar fields*, which are similar to electric or magnetic fields except that they have no direction. The Earth's magnetic field evidently has direction, for example, because it pulls the needle of a compass into alignment to the north magnetic pole. A weather map showing the temperature at various points around the country is a simplified example of a scalar field.

Exactly which scalar field caused inflation – many may have been around in the universe at the time – isn't clear. So, as a default, scientists refer to this key field as *the inflaton field*.

Guth's idea was that the scalar field at the very beginning of the universe was in a state called a *false vacuum*, which means that it was very dense with energy but was behaving as if its energy level was unable to go any lower.

Consider the following analogy to better understand the nature of a false vacuum: Imagine you're eating some peas. You bring a forkful up to your mouth but one falls off, rolling down your shirt-front until it gets caught in a fold of material halfway down. Bear with us here! The pea could go farther down, but something (the fold of shirt material) is stopping it, making the pea behave as if it can't go any farther.

Like the pea, the energy level of the inflaton field before inflation happened was suspended in a false vacuum state, ready to roll farther down at any moment.

In this false vacuum state, the inflaton exerted a large, negative pressure. According to Einstein's general theory of relativity, this pressure has the effect of creating a repulsive gravitational field, pushing away instead of pulling together. Thus a massive build-up of negative pressure was the force that drove inflation.

Guth's original idea was that as the universe cooled, the vacuum energy of the scalar field underwent a *phase transition*, a process similar to what happens when steam (water vapour) condenses into liquid water. When water moves between phases – for example, from gas to liquid – energy is released in a form known as *latent heat*. Similarly, when the scalar field changed phases, Guth said the field also released a huge amount of energy that caused the universe to expand rapidly.

According to Guth, the false vacuum decayed in 10^{-32} of a second and its energy was released, producing the hot mixture of particles with which the Big Bang theory begins, as we outline in Chapter 6.

Tweaking inflation

The trouble with Guth's original idea for inflation – as he acknowledged himself at the time – was that the phase transition he described doesn't end uniformly.

The physics here is complex, but fundamentally the problem is that bubbles of space would form where the inflaton field had decayed but the space around these bubbles had not. The end result of this situation is an unacceptable level of unevenness, or *inhomogenity* as scientists call it, in the universe.

Solving yet more problems

In addition to solving the smoothness, flatness, and horizon problems (see the section 'Solving some tricky problems'), inflation also offers a convincing explanation for how large scale structures like galaxy clusters evolved throughout the universe – another question unanswered by standard Big Bang models.

The link between inflation and galaxy formation starts with tiny fluctuations in the matter of the very early galaxy, called *quantum fluctuations.* These infinitesimal blips constantly appear and disappear throughout the universe without you noticing (see Chapter 9 for a discussion of Heisenberg's uncertainty principle, which allows energy fluctuations over short time scales). Inflationary theory says that when inflation happened, these vanishingly small fluctuations were captured and magnified enormously, persisting to become the fluctuations that scientists have seen in the cosmic background radiation (which we describe in Chapter 6).

As it happens, inflationary theory also makes some predictions about what those fluctuations should look like. Some of those predictions have recently been supported by measurements taken by the Wilkinson Microwave Anisotropy Probe.

Soon enough, though, some of the finest minds in cosmology got to work on overcoming the problems with Guth's model. Dozens of new versions of inflation were proposed, notably by scientists such as Andrei Linde, Andy Albrecht, and others. These variations include some models that don't require phase transitions at all.

Today, most cosmologists are convinced by the idea of inflation, although no one has proven beyond doubt that it even happened. Even so, the theory matches very well what scientists see in the universe.

Chapter 8

Thinking Differently About the Universe

In This Chapter

- Examining the Big Bang's big challenger – steady state theory
- Cruising through alternative cosmological theories – from plausible to fantastic

When you consider the huge size, age, and complexity of the universe, you can see how coming up with a theory that explains how everything came into existence has been difficult. Even today, cosmologists admit that the combined concepts of the Big Bang and inflation (refer to Chapters 6 and 7, respectively) don't explain everything to everyone's satisfaction.

Even though cosmology is a relatively young field of study, scientists have reached a point of virtually unanimous consensus: The theories of the Big Bang and inflation explain an awful lot about the cosmos and, as such, most professionals accept these theories to be the best currently available explanations for the origins of the universe. (This philosophical approach is similar throughout all scientific disciplines – achieve a consensus then keep testing and nibbling away to see whether the theory cracks under the scrutiny or someone comes up with a better idea.)

Still, a handful of doubters have proposed (and continue to propose) alternative scientific theories regarding the origins of the universe. In this chapter, we show some of the most popular and compelling competing ideas on how the universe was created (if indeed it was created rather than simply existing for all time).

Existing Forever: An Alternative to the Big Bang

Although the Big Bang may seem to be the only game in town today, this wasn't always the case. In the years after Hubble's observations, which seemed to show that the universe is expanding, not everyone was convinced of the idea of an initial explosion.

Sharing out the raisins: Conceptualising a constant universe

When trying to come up with models of how the universe was created, cosmologists often start off with various assumptions.

One of the most common starting points is that the universe is essentially the same wherever you look and in whichever direction you look. That may sound a little odd, given that you can see more stars and galaxies in some directions than in others.

However, when cosmologists say that the universe is the same they mean that when you consider the cosmos on a very large scale, the universe is pretty much the same: some galaxies, some stars, some dust, and some radiation.

Imagine that you were able to dice the universe up into huge cubes – perhaps trillions of light years across. If you then compared one cube to the next, you'd see a similar distribution of stars, galaxies, and other stuff. This situation is known as the *cosmological principle*.

Think of the cosmological principle like this: Imagine cutting a wedding fruit cake into cubes for guests at the reception. If you underestimate the number of guests and the cubes need to be really small, one guest may get a raisin and an almond in his cube while another may end up with a date and a walnut in her cube. If you have to serve fewer guests, you can cut larger cubes of cake and every guest probably gets one of every sort of fruit and nut in his or her cube.

The ideas behind the cosmological constant were used by Copernicus (who we talk about in Chapter 2) and others. They were also important in the calculations of Albert Einstein and other modern scientists. In fact, many leading theories of cosmology start off with the cosmological principle as their first assumption. Apart from anything else, it makes the sums easier (well, a little).

The cosmological principle also relegates the Earth to obscurity – and for this reason some people find it hard to accept. For most of history, people believed that the Earth was at the centre of everything. Only when Copernicus came along with his ideas of a Sun-centred (or *heliocentric*) universe did people start wondering whether the Earth was so important, and Kepler's work on planetary motion firmly quashed the Earth's special position. But even after Kepler, many people believed that the Earth's solar system or the Milky Way galaxy was at the centre of things. Hubble's observations (Chapter 5 goes into more detail on Hubble) meant that even these ideas had to be discarded. Thus for some people the cosmological principle says that no one has a special place in the universe and that the universe is pretty much the same everywhere.

Earth-bound humans truly are insignificant it seems – at least in cosmological terms. Of course, that doesn't stop some people insisting that the universe revolves around them!

Going steady: Steady state theory

Steady state theory – the theory that perhaps lays claim to being the most believable alternative to the Big Bang – was conceived in the 1940s. During the Second World War, the astronomer Fred Hoyle was, like many others, involved in war work rather than his day job of being a scientist. Hoyle headed a Royal Navy team of experts (including Austrian scientists Hermann Bondi and Tommy Gold) that was researching the recent innovation of radar. All three researchers also had an interest in the burgeoning science of cosmology.

In 1948, Hoyle, Bondi, and Gold produced two scientific papers laying out their ideas on a theory of the development of the universe, which they named steady state theory.

Steady state theory extends the idea of the cosmological principle through time. The three scientists suggested that not only is the universe the same wherever you look at it, but also that it's the same *whenever* you choose to look at it – 50 billion years ago, 50 seconds ago, or even 50 million years in the future. They didn't mean that the universe is unchanging – supernovae and the like disprove that idea – but instead steady state theory proposes that the *density* of the universe remains unchanged. That is, the average amount of matter in any given volume of the universe stays the same.

Considering the creation field

The challenge for Hoyle was that Edwin Hubble's red-shift measurements showed the galaxies moving farther apart. If this was the case then the universe wouldn't remain the same over time – it would become *less* dense.

Hoyle's solution was to come up with something called the *creation field,* an emotive name if ever there was one. Hoyle suggested that some process exists in which matter is constantly created, and this forms the basis for new galaxies to fill the space left empty as existing galaxies move outside the observable universe. (In the 1948 paper, Hoyle didn't speculate on the exact nature of this process.)

The nub of the paper is that the total density of matter in the observable universe (which we talk about in Chapter 7) remains constant through time. According to this model, the universe has no beginning or end in time, and conforms to the cosmological principle. Figure 8-1 helps explain this idea a bit more. The dotted line shows the observable universe. New matter is constantly created, which eventually coalesces into galaxies, keeping the density of the observable universe constant.

Hoyle's rationale for steady state theory was unhappiness with the idea that a Big Bang just happened for no apparent reason. In the introduction to his 1948 paper, Hoyle said that his theory handles

> *aesthetic objections to the creation of the universe in the remote past. For it is against the spirit of scientific enquiry to regard observable effects as arising from 'causes unknown to science', and this in principle is what creation-in-the-past implies.*

So why isn't steady state considered to be a viable theory for explaining the universe? Penzias and Wilson's discovery of cosmic microwave background (CMB) radiation in 1965 (described in Chapter 6) sounded its death knell. Steady state theory has no explanation for what caused CMB. In contrast, the idea of CMB being the afterglow of the Big Bang is both appealing and explainable in terms of that theory.

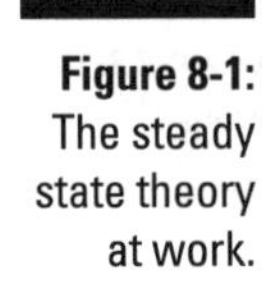
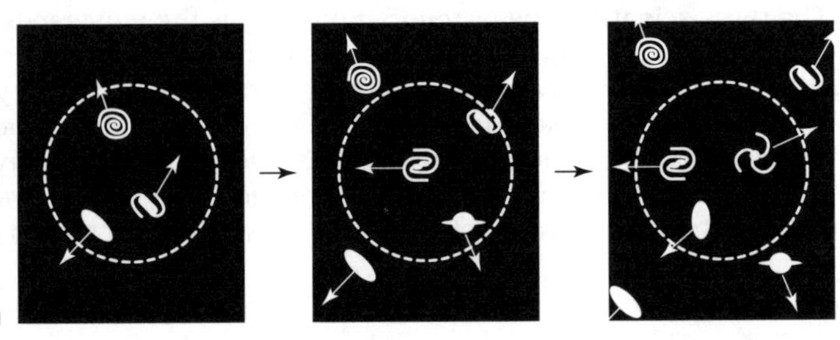

Figure 8-1: The steady state theory at work.

This failure, however, didn't stop Hoyle from being attached to steady state theory. He continued tweaking the theory, as we discuss in the section 'Evolving to the quasi stead state theory'.

Science – or science fiction?

Fred Hoyle was born in the middle of the First World War in Bingley in Yorkshire. He was always interested in maths and was apparently able to write out his times tables up to 12 × 12 by the age of 4, although he was unable to read until the age of 7.

According to his autobiography, he was a regular truant from his first school in Eldwick and found the chopping and changing between subjects every hour too much to bear.

Despite those behavioural issues, he managed to win a scholarship to Bingley Grammar School and went on to Emmanuel College, Cambridge, in 1933. His focus in his years at Cambridge was more on mathematics than cosmology, but he picked up an enthusiasm for the subject from his lecturers at the time. Hoyle was made a fellow of St John's College in 1939, just before the outbreak of war.

His wartime work put him in close touch with astronomers. His work with Hermann Bondi and Tommy Gold (see the section 'Going steady: Steady state theory') made him switch his focus to cosmology, in which he made several key contributions:

- **Accretion theory.** Hoyle, along with astronomers Hermann Bondi and Raymond Lyttelton, proposed the idea that stars form from collapsing clouds of gas. This theory, still credible today, says that these clouds collapse under gravity into rotating *accretion disks.* As more and more material concentrates towards the centre of the disk, the pressure and temperature grows to a point when nuclear reactions begin – and the star starts shining.
- **Interstellar dust.** Hoyle was a pioneer in studying the nature of the dust that litters the space between the stars and how it's formed during explosive events, such as novae and supernovae. This was a stellar contribution – literally and metaphorically – and we describe more about supernovae in Chapter 12.
- **Nucleosynthesis of the elements.** In 1957, with William Fowler of the California Institute of Technology and British astrophysicists Geoffrey and Margaret Burbidge, Hoyle described how the pressures and temperatures at the centre of a star can lead to the creation of elements beyond hydrogen. (Take a look at Chapter 12 to read more about this important concept.)

Hoyle also wrote science fiction. His most famous work, *The Black Cloud,* was published in 1957 and deals with the arrival in the solar system of an interstellar cloud, which turns out to be a living organism. Perhaps inevitably, Hoyle also strongly believed in the idea that life existed elsewhere in the universe.

Evolving to the quasi-steady state theory

Far from ditching the idea of steady state theory, Hoyle continued working on it throughout his career, seeing whether he was able to make modifications so that his central idea fit with newer observations, particularly the troublesome cosmic microwave background (CMB).

In 1993, Hoyle, Geoffrey Burbidge, and J.V. Narlikar from India's Inter-University Centre for Astronomy and Astrophysics proposed the *quasi-steady state theory.* As with steady state theory, this revision theory proposes that the universe has lasted and is going to last forever, but instead of mass being created regularly through the creation field, the three scientists proposed a series of 'little big bangs'. Quasi-steady state theory says that matter is created all the time at a low background rate but with occasional intermittent bursts involving lots of matter – around ten thousand trillion times the mass of the Sun each time.

This new theory has the benefits of explaining away the CMB by saying that it is a relic of the light from burned-out stars. (Of course, it faces the problem that scientists don't observe any such little bangs.)

In their conclusions, the scientists said:

> *This paper is not intended to give a finished view of cosmology. It is intended rather to open the door to a new view which at present is blocked by a fixation with Big Bang cosmology and, on a smaller scale, by an obsession with black holes.*

Other scientists have tried to find fault in quasi-steady state theory, and some argue that the predictions about the cosmic microwave background and other cosmic phenomena made by the theory don't exactly match the observations. However, these objections are unlikely to bury the quasi-steady state theory. Although Hoyle died in August 2001, other scientists continue to champion this alternative explanation of the universe.

Explaining the Universe in Other Ways

Other descriptions of the universe arise when scientists try to explain some of the baffling characteristics of the cosmos. For example, as we discuss in Chapter 7, scientists were originally baffled by the so-called 'horizon problem'.

The horizon problem is the cosmologist's way of referring to the fact that the cosmic microwave background (CMB) radiation has a constant temperature of 2.7 kelvin to better than 1 part in 10,000 everywhere, even places separated by distances between which light would have been unable to travel since the Big Bang.

The most widely accepted explanation for this is the concept of inflation. But other scientists had other ideas to explain this puzzle, as we explore in the following sections.

Combining cosmic ingredients: The Mixmaster Universe

In 1969, Charles W Misner of the University of Maryland proposed a theory about the origins of the universe called the *Mixmaster Universe.* Despite sounding like a compilation CD of classic house tracks, this theory in fact refers to the classic American kitchen appliance, the Sunbeam Mixmaster.

Misner's idea is that any irregularities in the early universe (that is, before the 'moment of last scattering' when light was first able to travel freely through the universe) are whisked out by a period of fluctuating chaos.

Misner came up with a solution to Einstein's equations, and theorised a universe that fluctuates between two shapes – one like a cigar, the other like a pancake. In Misner's theory, the universe expands in one direction while contracting in the other. Misner thought that wobbling between these two shapes smoothes out any inconsistencies and produces the CMB radiation that scientists observe today.

However, scientists have sidelined the Mixmaster Universe theory because

- The theory of inflation (flip to Chapter 7 for more on inflation) is more elegant and explains the same results more simply.
- The Mixmaster Universe theory runs into a problem with entropy. *Entropy* measures how disordered or chaotic a system is, and Misner's theory generates too much entropy compared with what scientists can observe in the current ordered universe.

Travelling through space: Tired light theory

Based on the work of Edwin Hubble and others in the first decades of the 20th century, scientists have explained the red-shift observed in light emitted from distant galaxies as being the result of the expansion of the universe (see Chapter 5 for details on how this happens). But the *tired light theory* offers another explanation. And from this theory, an alternative theory of the origins of the universe arises.

The originator of the tired light theory was Swiss astronomer Fritz Zwicky. He was a brilliant theoretician – the first to come up with the idea that you can deduce the existence of dark matter (see Chapter 11) from its gravitational effect on what is observable. He also coined the term *supernova* and was the first to suggest that supernovae are the origin of *cosmic rays,* the high energy particles that cascade down onto the Earth (and everywhere else, for that matter) from outer space.

Instead of accepting Hubble's view on red-shift, Zwicky came up with the idea that the light the Earth receives from distant galaxies has become 'tired' on its long journey to the Earth. By 'tired' Zwicky meant that the photons of electromagnetic radiation (Chapter 6 has more on electromagnetic radiation) lose energy in the intervening space because of gravitational interactions with massive objects along the way.

In Chapter 9, we look at quantum mechanics and discuss how light is made up of particles called photons, whose energy is directly proportional to the frequency of the light. The key importance for our discussion here is that as the energy of photons increases, so does the frequency of the light; as the energy of photons drops, the frequency follows. Based on this phenomenon, Zwicky suggested that the observed red-shift is just a sign of the photons from the distant galaxies losing energy, which he called *gravitational drag.*

Tired light has only ever had a small number of adherents, but the theory is still talked about from time to time as an alternative to the Big Bang by those who dislike the idea of an expanding universe. However, critics of tired light point out a number of key flaws:

- Gravitational drag is essentially a type of light scattering. When light is *scattered* (that is, bounced around in different directions because of its interactions with particles in the vicinity), its spectral lines are usually blurred because the process of scattering not only changes a particle's energy but also its momentum. However, scientists don't see this blurring when looking at light from distant galaxies.
- The light from supernovae in highly red-shifted galaxies appears to fade more slowly than supernovae in nearer galaxies. This is exactly what you'd expect to happen according to the standard understanding of the expanding universe. But tired light theory has no explanation for this observation.

Toying with matter and antimatter: Plasma cosmology

Another theory of the origins of the universe is based on the existence of four – not three – states of matter.

When you first start to study science at school, you're introduced to the three states of matter – solid, liquid, and gas. These days, however, this exclusive club of three has opened up membership to a fourth state – plasma.

Atoms are made up of a central core called the *nucleus* and a surrounding cloud of electrons. (We discuss atomic composition in greater detail in Chapter 9.) An atom's outer electrons can be stripped off by a number of different methods to leave a positively charged *ion*. When gas becomes *ionised,* the result is plasma – a cloud of positively charged ions and negatively charged electrons.

Plasmas tend to act differently from gas atoms that aren't ionised, hence the idea of a fourth state of matter. In fact, layers of plasma surround the Earth. One characteristic of plasmas is that they reflect radio waves. This reflectivity is why you can listen to medium wave radio stations even when you're out of the line of sight of the radio transmitter.

Why is all this talk of plasma important to the origins of the universe? Well, in the 1960s, physicists Hannes Alfvén and Oskar Klein came up with the idea of *plasma cosmology*. This theory proposes that the universe is made up of *ambiplasma,* a plasma comprised of both matter and antimatter in equal quantities (check out Chapter 9 for a discussion of antimatter). The theory states that this odd mixture can separate out into pockets of matter and antimatter. Between these pockets, matter and antimatter annihilate each other, producing vast releases of energy. This ambiplasma would be long-lived because, Alfvén argued, electromagnetic interactions seen in plasmas generated in the laboratory would stop this annihilation from happening too quickly. Other scientists aren't convinced that these electromagentic effects can be scaled up from the lab to the universe in this way.

One of the key attractions of plasma cosmology is that it may explain why an awful lot of matter seems to exist in the universe but not very much antimatter. Alfvén and Klein's answer is that humans live inside one of these pockets of matter. If things had been different, humans may have ended up in another pocket and may have been made out of antimatter. Heady stuff, isn't it?

Low level interest in plasma cosmology has existed since its conception, but most scientists now believe that the discovery of the cosmic microwave background (described in Chapter 6) puts paid to the theory. In addition, scientists theorise that the annihilation of so much matter and antimatter would give rise to large amounts of gamma radiation, far more than has ever been detected in the universe.

Exploring other oddities in cosmology theory

Cosmological theories don't stop with the Mixmaster Universe, tired light theory, and plasma cosmology. Have a quick shop around to see what else thinkers and scientists have come up with to explain the origins of the universe. Here are a few other theories to start with:

- **Modified Newtonian dynamics (MOND).** Some cosmologists who don't feel comfortable with Einstein's general theory of relativity (refer to Chapter 4 for more on Einstein), have instead turned to Newton's theories of gravity (described in Chapter 3) and argued that Newton's universal law of gravitation isn't universal at all but has different characteristics over very small and very large distances.
- **Rotating universe.** The logician Kurt Gödel came up with the idea of a *rotating universe,* which obeys Einstein's field equations and uses specific value for Einstein's cosmological constant (Chapter 6 has more details on this constant). In this model, the matter of the universe is literally rotating. It has some very odd characteristics, including the ability of objects to travel through time and for observers to see themselves at earlier times – if only they look in the right direction. Most cosmologists believe that Gödel's solution is a mathematical oddity rather than a realistic description of the universe.

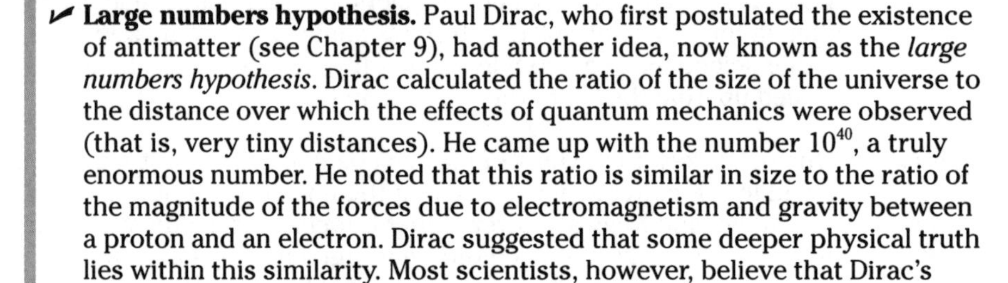

- **Large numbers hypothesis.** Paul Dirac, who first postulated the existence of antimatter (see Chapter 9), had another idea, now known as the *large numbers hypothesis*. Dirac calculated the ratio of the size of the universe to the distance over which the effects of quantum mechanics were observed (that is, very tiny distances). He came up with the number 10^{40}, a truly enormous number. He noted that this ratio is similar in size to the ratio of the magnitude of the forces due to electromagnetism and gravity between a proton and an electron. Dirac suggested that some deeper physical truth lies within this similarity. Most scientists, however, believe that Dirac's mathematical similarities are just a coincidence.

Part III
Building Your Own Universe

'Isn't that the new Australian astronomer?'

In this part . . .

In this part, we play with the basic building blocks of the universe, the minuscule particles that make up everything we see around us. These building blocks need some form of cosmic glue to keep them together, so dive into this part to discover the four fundamental forces of the universe.

You may think that the dark side is something that only exists in the Star Wars universe. Well, our own universe has its dark side too in the form of dark matter and dark energy. Delve into this part to find out more.

The science of chemistry – the link between the unseen subatomic realm and our everyday world – is given the spotlight in this part. Did you know that everyone is made of stardust? That's cool, but how did that jumble of chemical elements formed by stars turn into a living thing? We try to answer that tough question in this part.

Finally, put on your hard hat and indulge in a bit of construction work – we build an entire universe from scratch, starting from stars and planets and ending up with the huge structures that span the cosmos.

Chapter 9

Building Things from Scratch

In This Chapter

- Creating the stuff of the universe, bit by bit
- Identifying gaps in the current understanding of matter
- Seeking strange subatomic things with particle physicists
- Tracking the ghost particle

The universe seems to be an astonishingly diverse place, full of weird and wonderful things. Consider the galaxies: These mega-objects can contain billions of stars of differing sizes and compositions. Around each of these billions of stars may circle planetary systems like our solar system, comprising planets as diverse as gassy Jupiter and rocky Mercury and other possible types of planet that scientists have yet to discover. And you can add to that mixture a wealth of strange objects, such as black holes and nebulous clouds of dust.

Adding to this diversity, consider the myriad life forms that inhabit the Earth and – who knows? – maybe elsewhere. At one end of the scale you have the huge mammals, such as the blue whale – incredibly complex organisms that have reached their current form through eons of evolution. At the other end, you have minute organisms such as bacteria and viruses. Although they may be small, they too are incredibly complex.

All told, the universe is a varied place to live. But perhaps the strangest thing about the universe is that all the wonderful variety it contains is made from the same basic materials.

The search for the basic ingredients of the universe has been going on for centuries. Some scientists argue that humans can never find an answer to what makes an atom function, because as each layer is peeled back, another layer is revealed within – like some sort of cosmic onion.

Still, this fact doesn't mean that the search for universal truth is fruitless. As scientists peel back the layers, they realise that they're looking farther and farther back in time, to the very particles that existed at the origins of the universe. Ironic, isn't it, that only by finding out about the smallest objects in the universe can humans discover how the universe itself has come to look as it does today.

What's the Matter? Searching for the Most Basic Building Block

Some 2,500 years ago, the ancient Greeks began thinking about the constituents of *matter*, a broad term covering everything we can touch around us, whether solid (like a table), liquid (like wine), or gas (like the steam rising off a hot bath). They decided that some basic building block from which everything was constructed must exist. For 2,000 years, their view of the atom as the fundamental object upon which everything else was built (even though they hadn't actually seen an atom or defined what it was) remained unchallenged.

Yet the search for basic truths about the universe made engineers and scientists in the 19th century question what atoms themselves were composed of. As researchers probed deeper inside atoms, they revealed the atom's more fundamental components – the electron, proton, and neutron. Perhaps these were the basic building blocks of the universe?

Getting smaller and smaller – but only to a point

Imagine that you're celebrating your birthday and you have a nice cheesy pizza to share with your party guests. If four friends come, dividing up the pizza is simple. If you ask everyone in your class or office and 30 turn up, dividing the pizza is a bit trickier but still possible. If everyone from your whole school or company turns up, you may still be able to manage, given a sharp enough knife (and guests who'll be satisfied with the world's thinnest slice). But how far can you go on chopping?

The ancient Greeks thought along similar lines (although history doesn't tell how popular pizza was at the time). In fact, the Greek philosophers Leucippus and Democritus, in the fifth century BC, came up with the idea that at some stage in cutting this theoretical pizza, you reach a point where you simply can't divide it any further – not because your knife isn't sharp enough but because matter is composed of fundamental, indivisible elements. (See the sidebar 'Pondering Plato's beautiful solids' for more on his take on the nature of matter.)

The Greek word for indivisible was *atomos*, which serves as the root for the English word 'atom'. Leucippus and Democritus believed that all objects were made from a combination of atoms and empty space.

Of course, the ancient Greeks didn't have the benefit of electron microscopes and had only the most rudimentary understanding of chemistry, so they were never able to perceive these atoms themselves. Another two millennia needed to pass before the Greeks' ideas were confirmed.

Pondering Plato's beautiful solids

Plato, who came a couple of centuries after Democritus and Leucippus (see the earlier section 'Getting smaller and smaller – but only to a point') had his own views of the universe and its basic building blocks.

In his philosophical work *Timaeus*, Plato explains how God created the universe from four elements: earth, air, water, and fire. Each of these four elements is formed from tiny and invisible geometrical solids: the earth from cubes, the air from 8-sided shapes called octahedrons, fire from pyramids, and water from 20-sided objects called icosahedrons.

Plato was convinced that the faces of these solids were made up of two types of right-angled triangle – 'one isosceles, the other having the square of the longer side equal to three times the square of the lesser side'.

These triangles and solids were the most beautiful and divine arrangement Plato was able to contemplate, and in *Timaeus*, he states:

> *For we shall not be willing to allow that there are any distinct kinds of visible bodies fairer than these. Wherefore we must endeavour to construct the four forms of bodies which excel in beauty, and secure the right to say that we have sufficiently apprehended their nature.*

Nice idea, Plato, but unfortunately wrong.

Discovering the electron: Thompson

The Greek concept of the atom's indivisibility endured until the 19th century when the work of several scientists began to question the behaviour – and ultimately the composition – of the atom.

Although many scientists in the 19th century experimented with electromagnetic energy and attempted to explain its nature, Manchester-born physicist Joseph John Thomson is generally credited with discovering the first of the particles that make up the atom – the electron.

He did so using various types of vacuum tube and cathode rays.

Ask any schoolchild a few years ago about cathode rays and they would have almost certainly known it was something to do with television. Before the arrival of flatscreen, plasma, and LCD screens, the world watched bulky television sets at the heart of which were cathode ray tubes. A cathode ray tube is essentially a glass bulb out of which all the air has been removed. Televisions worked by projecting a stream of electrons – the inaccurately named 'ray' – through the tube and onto a special screen.

But at the time that Thompson was doing his experiments, electrons were unknown to science, and cathode tubes were a mystery. He was able to show in a string of cleverly designed experiments that when an electrical current was passed through a cathode tube, the current was carried by negatively charged particles, which he called 'corpuscles'. In his paper on his experiments, he says:

> *As the cathode rays carry a charge of negative electricity, are deflected by an electrostatic force as if they were negatively electrified, and are acted on by a magnetic force in just the way in which this force would act on a negatively electrified body moving along the path of these rays, I can see no escape from the conclusion that they are charges of negative electricity carried by particles of matter.*

Instead of corpuscles, scientists today call the charged particles that Thompson worked with *electrons*. This name comes from Irish physicist George Johnstone Stoney. Stoney's work on how gases move around (so-called *kinetic theory*) led him to name the basic unit of electricity as the electron, after the Greek word for amber (the Greeks knew that rubbing amber with fur created sparks).

Dissecting the atom further: Ernest Rutherford

The discovery of the electron was the catalyst for scientists to study the basic constituents of matter more closely than ever before.

Ernest Rutherford was born in Nelson, New Zealand in 1871. He studied at Canterbury College at the University of New Zealand before heading half way around the world to do postgraduate research at the Cavendish Laboratory in Cambridge, England. Like many of his colleagues, Rutherford was interested in electricity but he went on to study the newly emerging field of radioactivity.

Radioactivity is about more than atomic bombs. In general terms, *radioactivity* is the word scientists use to describe the different processes by which certain unstable atoms can spontaneously emit energy and particles.

Rutherford's discoveries were fundamental in starting to understand the universe and included:

- **Identifying three forms of radioactivity.** By looking closely at radioactive uranium, Rutherford showed that more than one kind of radioactivity exists. In fact, he found three:
  - Alpha rays or particles (essentially atoms of helium that have had their electrons stripped off)
  - Beta rays (streams of electrons)
  - Gamma rays (extremely energetic radiation)
- **Conceptualising a basic model of atomic structure.** When Rutherford aimed alpha rays directly at a thin sheet of gold foil, the alpha rays usually passed through the foil. However, sometimes they bounced off to the side or even bounced back towards the source (see Figure 9-1). A is a particle that passes through with no deflection, B is a particle that passes through the foil but is deflected off course a little, and C is a particle that has bounced back after hitting a gold atom.

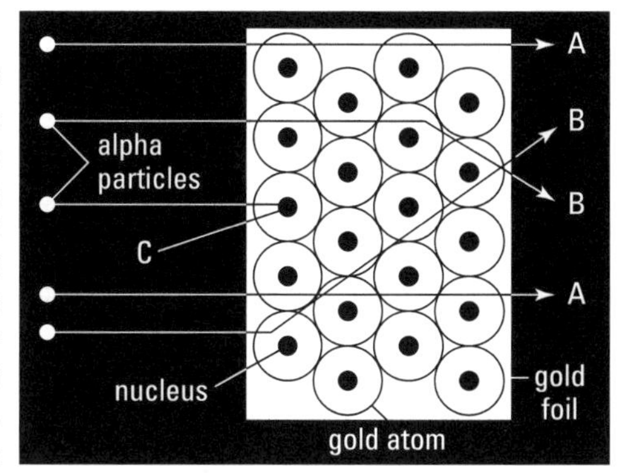

Figure 9-1: Alpha rays directed at a thin sheet of gold foil either pass through or bounce off the foil.

Rutherford conceptualised that the gold foil was made of many atoms, each having a compact central nucleus that contains most of the atom's mass and all its positive charge. The area around the nuclei was pretty much empty space with the odd electron flying around. When alpha rays encountered the space outside a nucleus, they proceeded through the foil. But when alpha rays encountered a nucleus they bounced back, first because the nucleus of the gold atom was far more massive than the alpha ray, and second because of the electromagnetic repulsion (both projectile and target were positively charged). Rutherford's basic model of atomic structure is still accepted today.

Noticing the neutron

Ever since John Dalton's work on atomic theory in the early 1800s and Dmitri Mendeleev's creation of the table of elements in 1869, scientists had realised that the elements could be arranged according to their increasing weight, or to put it more accurately their *atomic mass.*

Ernest Rutherford credited the increasing mass of elements in the periodic chart to an increase in the number of protons in the nucleus. For example, the atomic mass of helium is around 4 relative to oxygen which was defined as having an atomic mass of 16. Hydrogen has an atomic mass of just over 1.

Rutherford speculated that a helium atom had a nucleus of four protons (thus giving it the correct atomic mass). He further hypothesised that helium had two electrons that were held together in some unknown manner and then two additional electrons that orbited some distance even farther out from the nucleus. This model of the helium atom was electrically neutral (four positive protons balanced by four negative electrons) and fit with the observed atomic mass of helium. Rutherford further postulated that each of the two electrons in helium were somehow bound together with a proton, forming something he called the neutron.

Unfortunately for Rutherford, the neutron eventually discovered by James Chadwick in 1932 isn't a combination of a proton and electron – it's a particle in its own right.

To identify the neutron, Chadwick bombarded a target made from the element beryllium with alpha particles. Other scientists had previously conducted similar experiments and produced highly penetrating 'radiation'. For a variety of reasons, Chadwick disagreed. He concluded that the 'radiation' was not radiation at all but a stream of highly energetic, electrically neutral particles – *neutrons* – with a mass almost identical to that of the proton.

- **Evaluating radioactive decay rates.** Rutherford realised that each radioactive element decays at a fixed rate. He came up with the concept of *half-life*, the amount of time required for half the amount of a radioactive element to decay into something else. For some elements this time is a fraction of a second, for others it's measured in billions of years.

  Rutherford's concept of half-life led to a re-evaluation of the age of the Earth itself. Scientists began re-thinking creation with the work of Darwin in the mid-1800s and now had further supporting evidence that the Earth was billions (not thousands) of years old. By measuring the concentrations of the products of radioactive decay, Rutherford was able to show that the Earth is by necessity billions of years old, long enough for evolution to take place as Darwin outlined.

- **Transmuting elements.** Alchemists had for centuries been trying to convert cheap metals into gold. Rutherford didn't do exactly that but he was the first to transmute one element into another.

  By firing alpha particles at nitrogen gas, he managed to convert some of the nitrogen with oxygen. Understanding the conversion of one element into another was an essential step in explaining the life cycle of stars, as we discuss in Chapter 12.

- **Discovering and naming the proton.** By bombarding a non-radioactive element with alpha particles, Rutherford managed to dislodge a single positively charged particle, which he named the *proton*.

In particular, Rutherford's model of the atomic nucleus and the discovery of the proton earned him the title of father of nuclear physics. He eventually became director of the Cavendish Laboratory and the likes of Niels Bohr, James Chadwick, and Robert Oppenheimer came to work with this creative genius.

Venturing Beyond Electrons, Protons, and Neutrons: Quantum Mechanics

In 1911, Ernest Rutherford came up with a good model for how the atom worked – a central nucleus that contains most of an atom's mass and all its positive charge in the form of protons, with electrons orbiting it like planets around a sun (see Figure 9-2). But just two years later, in 1913, the model changed again because the atom's very small structures don't fit the rules of classical mechanics.

Based on *classical mechanics* (the system of laws, based in large part on Newton's work, which govern the predictable motion of everyday objects) and Maxwell's equations of electromagnetism (refer to Chapter 4 for more on Maxwell), an atom's electrons continually emit radiation, and so therefore should lose energy and eventually spiral down into the nucleus.

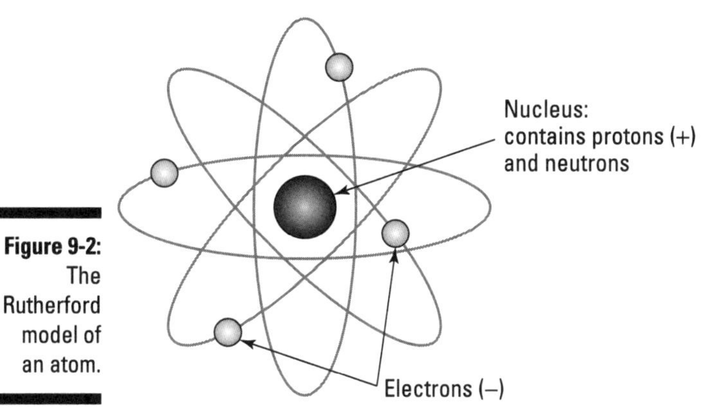

Figure 9-2: The Rutherford model of an atom.

This disastrous result doesn't come about, however, which was a good clue that scientists needed another way to think about the working of atoms. That new framework has become known as *quantum mechanics* – the science that describes how things work on the scale of atoms and their component parts.

The birth of quantum mechanics: Max Planck

Max Planck took the first steps to the new science of quantum mechanics. Planck was born in Kiel, Germany, in 1858 and studied under two of the leading figures of the growing field of *thermodynamics* (the study of changes in heat and energy), Hermann von Helmholtz and Gustav Kirchhoff.

Planck was interested in *blackbody radiation*, which we discuss in Chapter 6. *Black bodies* are objects (not necessarily black in colour) that absorb all the radiation that falls on them and then re-emit the radiation at all possible wavelengths. The spectrum of radiation these black bodies produce has a distinctive shape, as Figure 9-3 shows.

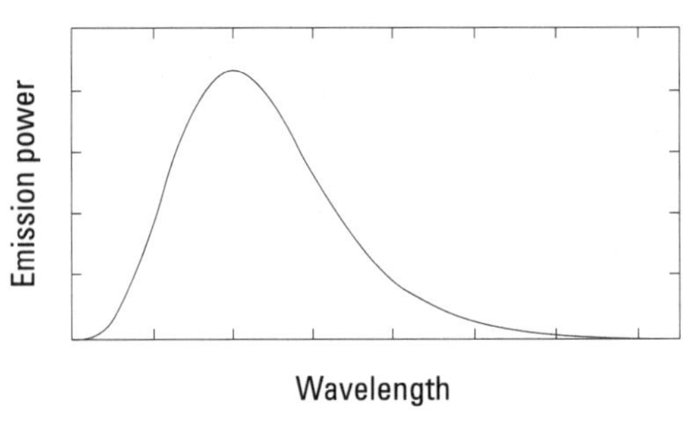

Figure 9-3: The radiation spectrum of a black body.

Planck realised that the shape of a blackbody radiation spectrum isn't consistent with classical mechanics, which predicted that the intensity of the radiated energy increases as the frequency increases. The problem with this idea was that as you consider higher and higher frequencies of radiation, the amount of energy radiated would approach infinity, which clearly doesn't happen. (This problem eventually became known as the *ultraviolet catastrophe* because ultraviolet light has high frequencies and the observed spectrum didn't fit existing classically derived predictions.)

In his 1900 paper 'On the Law of Distribution of Energy in the Normal Spectrum', Planck argued that the radiated energy from black bodies could only be emitted in finite 'bundles' of energy. This division of energy into packets with a certain value for the energy (which Planck established during his research) is called *quantisation*; each individual packet of energy is known as *quantum*.

This discovery means that light isn't a continuous phenomenon; instead it's made up of these little bundles or packets of energy that can't be broken down into any smaller amounts. It also means that rules must govern the ways these packets can be combined to make radiation.

This may be hard to imagine but just consider for a moment what would happen if the Queen suddenly decided that she didn't like certain coins of the realm and that, henceforward, only 10p, 20p, and 50p coins would be legal tender. Planck's discovery was something similar to this situation.

Suddenly, shopkeepers would only be able to use certain prices for the things in their shops. Pricing items at 7p, £2.26, and £4.99 would be impossible because smaller coins were no longer available. In fact, only prices ending with a round 10p would be possible rather than the full range of prices available before.

Planck didn't suggest how quantisation may actually work in reality. The entire concept was merely a mathematical contrivance that fit his observations.

Still, his work had an enormous impact on how scientists thought about the atom – and was one of the building blocks for quantum mechanics.

Conceptualising the atom – again

Scientists have long known that when certain gases are heated up or otherwise bombarded with energy they emit specific colours of light. An everyday example is the sodium streetlamp, which when turned on casts a distinctive pinky-orange colour over the road. If the Rutherford model of the atom were correct, these distinctive colours wouldn't be observed and white light (made up of a mixture of different colours) would be emitted.

Niels Bohr, a Danish physicist, considered how quantisation may explain the difference between what scientists saw in real life and what Rutherford's model of the atom suggested. He argued that rather than the electrons whizzing around in orbits with any amount of energy and at any distance from the central nucleus, they could only be in specific orbits, with specific energies, and at specific distances from the nucleus.

Bohr proposed that electrons didn't gradually lose energy, as classical mechanics predicted. Instead, an electron could only fall or jump between allowed orbits if it changed its energy by exactly the difference between its current energy level and that of another permitted orbit. If the electron fell to a lower orbit, it would have to get rid of this energy in some way, which Bohr proposed would be through giving off a single *quantum* (a bundle of energy) of electromagnetic radiation (known as a *photon*) of that exact amount of energy. Conversely, if an electron were hit by a photon with just the right amount of energy, it could jump up to a higher energy level or orbit.

This idea would explain the distinctive colours emitted by gases. Every atom of sodium can only have electrons with these well-defined energy levels and the photon of radiation emitted has a fixed energy (the difference between two fixed energy levels). A photon of light that has a fixed energy has a fixed wavelength and it is the wavelength of light that gives it a distinctive colour.

Travelling in waves or particles – or both?

Ever since the 16th century, the majority of scientists had believed that light travelled in waves, just like the ripples on the surface of water.

A number of experiments backed up this idea. The most famous showed that when you shine light through two narrow slits, the light on a wall beyond the slits forms bands of light and dark. These bands are similar to the large peaks and troughs of water formed when two sets of ripples meet in a pond (see Figure 9-4).

However, not everyone shared this view. Notable among them was Isaac Newton who believed in a 'corpuscular' theory of light, in which light was made up of tiny particles.

As it turns out Newton was right, according to a certain youngster named Albert Einstein.

Einstein studied something called the *photoelectric effect*, in which some forms of light or radiation when shone on certain types of metal caused electrons to break free from the metal's surface. This effect can only be satisfactorily explained if the light is made up of photons. Some of this photon energy is used to overcome the electromagnetic attraction keeping the electron within the metal.

Figure 9-4: Shining a light through two narrow slits (A and B) to make bands of light and dark.

Einstein won the 1921 Nobel Prize for this work on photoelectric effect, not for his theories of relativity.

Making a compromise: Karl Heisenberg

So on one hand, numerous experiments showed that light behaves like a wave, whereas Einstein's explanation of the photoelectric effect proved that light is made up of particles. This strange state of affairs is known as *wave-particle duality* and is one of the cornerstones of quantum mechanics.

This duality is not restricted to electromagnetic radiation. Shortly afterwards, electrons were shown to behave in a manner just like light – as both a wave and as particles. In fact, if the theorists are to be believed, every object in the universe behaves as both a wave and a particle, even you and me. However, the more massive the object, the smaller the scale over which this wave-like nature can be seen. For most everyday objects, the wavelength involved is so small as to be impossible to perceive.

Karl Heisenberg was born in Bavaria, Germany in 1901 and entered the University of Munich in 1920. Despite wanting to study pure mathematics, he was dissuaded after a poor interview and instead started a degree in theoretical physics. He achieved his doctorate just three years later and at the young age of 25 was made a professor of theoretical physics at the University of Leipzig.

In 1926, while working as an assistant to Max Born, Heisenberg imagined a new piece of scientific equipment called a *gamma ray microscope*, which differed from normal microscopes.

Microscopes in a typical school biology lab use reflected light to operate. An object is placed beneath a glass slide and the light shining on it is reflected through a system of lenses into the eye of the observer.

By contrast, Heisenberg imagined using gamma rays instead of visible light in his innovative microscope. The reason? Classical optical theory says that a lens's *resolving power* – its ability to distinguish small features in an object under scrutiny – is better the shorter the wavelength (and the higher the frequency) of the light used to illuminate it. And gamma rays have very short wavelengths.

Heisenberg had no intention of making a gamma ray microscope but wanted to imagine what would happen if you used one to look at subatomic particles. Heisenberg imagined shining gamma rays onto a single electron under this microscope and seeing what it looked like. The problem that Heisenberg realised is that because of wave-particle duality (see the preceding section 'Travelling in waves or particles – or both?'), the gamma photon would whizz in and hit the electron, knocking it from its position at the centre of the microscope's field of view, effectively meaning you couldn't observe the electron any more.

Heisenberg later realised that in subatomic situations, measuring both the position and the momentum of an electron (or any other particle for that matter) with complete accuracy is impossible. Equally impossible is measuring both the energy and the time that the particle had that energy with perfect accuracy. The idea that you have to compromise – you can measure one physical quantity with great accuracy but only at the expense of knowing another closely-related physical quantity poorly – has since become known as *Heisenberg's uncertainty principle*.

The uncertainty principle blurs the concepts of position and time. In classical mechanics, everything about an object's position and path is well defined. Knowing an object's position at some particular time and the forces acting upon it, you can work out with precision where it's going to be at some time in the future.

In quantum mechanics, however, things are a lot less clear. Because of the uncertainty of momentum and position, you can't say exactly where a particle will be after an interaction. You can only assign a *probability* that it will be in a certain position or have a certain momentum. Scientists describe that probability using fiendish maths which we steer clear of here.

In any case, don't feel bad if the whole concept of quantum mechanics strikes you as strange, counter-intuitive, and frankly bizarre. Many of the world's greatest scientists have felt, and feel, the same way. But quantum mechanics really does describe the world (and the universe!) fantastically well.

Probing the Concept of Probability

In 1926, Austrian physicist Erwin Schrödinger was working at the University of Zurich in Switzerland. Schrödinger didn't like Bohr's model of the atom (see the earlier section, 'Conceptualising the atom – again') and came up with a model based on energy moving as waves, not unlike waves moving on the surface of a body of water.

Schrödinger's wave equation proved spectacularly successful at predicting the observed atomic spectra of a hydrogen atom. The question then arose as to what this equation meant in reality.

With a wave equation governing the ripples on a pond, the answer is easy – the equation dictates where the individual drops of water on the surface are at any given moment in time. But the same can't be true for an electron orbiting inside an atom because this would mean that the electron is somehow smeared out over the entire orbit.

It was Max Born, Heisenberg's professor at the University of Leipzig, who came up with an interpretation that has since been widely adopted, despite flying in the face of common sense. Born argued that Schrödinger's wave equation indicated the probability of finding the electron in a particular location rather than dictating the position itself.

This idea means that instead of saying an electron is in a particular spot, all you can do is plot a cloud of points around the nucleus where it possibly may be. In some senses, you can think of that crazy electron being anywhere and everywhere simultaneously, except that it's much more probable that you're going to find it in some places than in others.

Born's interpretation of the atom gives rise to one of the strangest features of quantum mechanics – *tunnelling*. Here's an example to illustrate tunnelling. Imagine that you're in your garden and suddenly a ball appears over the fence from next door. Pretend that you've never been very good at throwing, but you try to throw it back over. Unfortunately, your efforts are so feeble that even when you throw the ball the hardest you can, it still doesn't manage to clear the fence.

The path of the ball is easily describable in the world of classical mechanics. For it to sail over the fence, you need to throw the ball with enough energy so that by the time it reaches the top of its trajectory (when gravity brings the ball to a halt), the ball must be higher than the top of the fence. If you can't generate that energy with your throw, the boy next door's football game is at an end (although he'll be rather amused at your wimpy throwing attempts by this stage).

By contrast, in the world of quantum mechanics, things are much stranger. The trajectory of the ball extends through the entire universe. At each point in the universe, you can calculate a probability that the ball can be found there. Therefore, the possibility exists (although it's very improbable) that the ball may be found on the boy's side of the fence – and as such, the situation is as if the ball has somehow tunnelled through the fence.

This may sound unlikely, but tunnelling happens in the real world. We talk in the earlier section 'Dissecting the atom further: Ernest Rutherford' about radioactive elements releasing *alpha radiation,* which involves the spontaneous emission of a bundle of two protons and two neutrons from the nucleus. Normally, the particles in the nucleus are bound tightly together and don't have enough energy to overcome their attraction.

Yet quantum tunnelling allows them to overcome this attraction every now and again; physicists have seen it in action in the laboratory. It's a strange old world indeed.

Antimatter

So, quantum physicists think of energy moving as waves in order to describe how electrons move around the nucleus of an atom. This model doesn't just help explain ordinary matter, however. It also led scientists to the discovery of antimatter.

Antimatter is one of the staples of science fiction novels and movies because when antimatter meets matter – as it did at the beginning of the universe and in carefully controlled lab experiments today – it annihilates in a huge burst of energy. Want to create a doomsday device? Simple, just suspend a dollop of antimatter in a vacuum and then send it into orbit around your enemy's planet. Right?

But what is antimatter? To understand it properly, we need to have chat about the weather, and talk about negative numbers.

Knowing the nuances of negative numbers

One of the only places that most people come into contact with negative numbers is the world of weather, where 'minus five' is five degrees below zero. However, mathematicians and scientists use negative numbers all the time and knowing how to use them in arithmetical operations such as multiplication is vital.

You've probably heard the saying, 'Two negatives don't make a positive.' That may be true in the world of childhood naughtiness, but the opposite is true in the world of maths – multiply two negative numbers and you get a positive one. Multiply –5 by –6 and the answer is 30, for example, the same as if you multiplied 5 by 6. Similarly, -9×-9 is 81, just the same as 9×9. (So the next time someone asks you what the square root of 81 is, don't just say 9, say 9 or –9 – although expect sneers of derision from anyone listening.)

So why is this important in the world of atoms and the sub-atomic particles they're made of? A physicist called Paul Dirac applied Schrödinger's wave equations to the electron and realised that they gave both positive and negative answers. As well as positive answers that describe everyday matter, Dirac found that the negative answers referred to something new and unexpected – antimatter.

Discovering the positive electron

In 1933, a physicist at Caltech, Carl Anderson, was using a special device called a Wilson cloud chamber (see the later sidebar 'Gazing at the clouds: The

Wilson cloud chamber' for more details). With the aid of the device, Anderson was watching the tracks left behind in the path of cosmic rays. Cosmic rays are streams of extremely high energy particles that are believed to originate from supernovae (see Chapter 11) and which permeate the universe. As a result, cosmic rays from outer space are constantly bombarding the Earth.

Anderson took 1,300 photographs of these tracks as cosmic rays passed through his cloud chamber and noticed something unusual in 15 of them. He considered the tracks

> *to be interpretable only on the basis of the existence in this case of a particle carrying a positive charge but having a mass of the same order of magnitude as that normally possessed by a free negative electron.*

In the clearest of Anderson's photographs, the cloud chamber's lead plate is visible across the middle of the photo and has the effect of slowing down any particles that pass through the chamber.

The curvature of the paths of particles in magnetic fields depends on their mass and velocity. Fast-moving particles have almost straight tracks because they pass through the chamber quickly. Heavy particles do as well – because of their mass, they are less affected by the magnetic field. By contrast, slow-moving and light particles have their tracks curved dramatically.

By studying the direction and amount of curvature, Anderson determined that the particle was positively charged and had the same charge as a proton. Anderson's paper presenting this finding was called 'The Positive Electron' and in it he coined the term *positron.*

Anderson's identification of the positron was the first direct observation of antimatter. The particle seems exactly like an electron, certainly as far as its mass is concerned, but is crucially different in that it's positively rather than negatively charged.

So what is antimatter?

Although the positron was the first antimatter particle discovered, every particle turns out to have its own 'evil' twin.

- **Particles and their anti-particles have the same mass.** Both the electron and positron have the same mass, for example.
- **Particles and their anti-particles possess identical *spins.*** Spin is a measure of angular momentum, or how difficult it is to stop something rotating, like a child's spinning top. The modern view is that fundamental particles don't actually rotate but interact as if they did (that is, spin

Gazing at the clouds: Wilson cloud chambers

Particle physicists, those scientists who specialise in the murky world of objects smaller than the atom, use a number of novel devices in their studies. In the early days of particle physics, one of the most commonly used devices was the cloud chamber.

The Scottish physicist Charles Wilson, inspired by a visit to the summit of Scotland's highest mountain Ben Nevis in the summer of 1894, invented the cloud chamber. While there, he experienced the atmospheric phenomenon known as the *Brocken Spectre,* where shadows fall onto a bank of mist and are surrounded by coloured halos. Wilson recreated a similar effect in the laboratory using a chamber equipped with a tight-fitting piston and filled with moist, dust-free air. By rapidly raising the piston, the enclosed air quickly expanded, bringing it close to the point where the liquid present in the air starts to condense. This process is similar to rain clouds forming in the atmosphere.

Wilson used the chamber to show how X-rays passed through the chamber made the air conductive. At the time, he didn't fully understand that the energetic X-rays were knocking electrons off the molecules in the air. Knocking an electron off an electrically neutral molecule turns it into a positively charged ion.

When the air in the chamber is just at its condensation point, these ions act as centres for condensation to build up. The path of an X-ray through the chamber thus appears as a track of condensation, which resembles wispy clouds – hence the chamber's name.

In 1911, Wilson had developed his cloud chamber to the point where he was able to view and photograph the paths of alpha particles and beta rays from radioactive sources. It was effectively the beginning of modern particle physics.

needs to be conserved in particle interactions). Odd but true. And as it turns out, particles and their anti-particles have identical spins. The electron and positron both have a spin of ½.

- **Particles and their anti-particles have opposite electrical charges.** The electron has a negative charge, whereas the positron has an equal and opposite positive one.

More exotic fundamental particles also exist, which have quantum mechanical properties odder than things like charge and spin. Some particles exhibit something called *strangeness*. The thing to remember is that the antimatter counterparts of these exotic particles always have opposite quantum properties, meaning that particle–anti-particle pairs can be created from energy without breaking any universal conservation laws and can then, just as quickly, disappear into a puff of energy. This situation is the crux of sci-fi's love of antimatter. Any particle meeting its anti-particle annihilates to produce a burst of energy.

If that's the case, you may be asking why we're still here. If every particle annihilates with its anti-particle, the universe should long ago have annihilated into a big pool of energy. However, it turns out that an imbalance between matter and antimatter exists, for which we should perhaps be rather thankful.

How particle accelerators work

We discovered from Einstein's famous equation (described in Chapter 4) that mass and energy are interchangeable. A lot of people think only of the destructive side of this – such as the creation of energy from the radioactive material in a nuclear weapon.

Yet the process can go in reverse and mass can be created from energy, and that's what happens in a so-called *particle accelerator* or *collider*. Most accelerators incorporate a long vacuum tube, usually circular or straight. Particles are accelerated along this tube using very strong electric and magnetic fields and then smashed into fixed targets or into other particles travelling in the opposite direction.

You may think that you'd only get the same particles out as you put in (albeit maybe a little dented!), but $E = mc^2$ means that the *kinetic energy* (the energy due to the motion of the accelerated particles) can be converted into mass too. The only caveat is that the total energy must balance before and after the collision.

So if you smash an electron into a positron at high speeds, sufficient energy is available to create new particles with higher masses than either of the original particles. Often these new particles are unstable and only hang around for a fraction of a second before transforming into other, lighter and more stable particles.

What use is that? Well, many people question the billions of pounds spent on experiments like the Large Hadron Collider at CERN, the European Particle Physics Laboratory, in Geneva. But the argument is that by probing these high energy environments we're looking back in time to the beginning of the universe when energies such as these were commonplace and the world looked a lot different than it does today.

The trouble is that when particle physicists started using particle accelerators, they realised that far more sub-atomic particles exist than they ever imagined.

Getting to Know the Standard Model

The 1940s and 1950s were an exciting time in particle physics. New and exotic particles were being discovered all the time, using cloud chambers, pressurised chambers full of liquid known as bubble chambers, and even photographic plates.

1947 was a particularly good year. Particle physicists in Bristol discovered particles called *pions* in the tracks left by cosmic rays in photographic plates, and researchers in Manchester found heavy electrically neutral particles that decayed into pions the same year. (These latter particles became known as *kaons*.)

Faced with all these new particles, one eminent particle physicists of the era, Enrico Fermi, said: 'Had I foreseen that, I would have gone into botany.'

With the discovery of dozens of new particles, physicists began to wonder whether these particles were actually made up of something even smaller and more fundamental. The answer was yes, as we shall discover later in this section. The particles that they discovered, which are now believed to be truly the most fundamental particles in existence (that is, they have no inner structure), together with the particles that are involved in interactions between these ultimate building blocks, are collectively known as the *Standard Model.*

The Standard Model is the best theory that scientists have come up with to explain how the things we see around us are constructed on the most fundamental level. The science involved can be fiendishly difficult to grasp, even for experts in the subject. However, this section gives you a good idea of just what those boffins get up to in their underground laboratories and particle accelerators.

Classifying quarks

In the 1950s, scientists started to be able to control beams of electrons and fired them at protons. By observing what happened to the electrons, scientists realised that all the charge and mass of a proton isn't concentrated at some central point but distributed among numerous smaller particles within the nucleus.

In the early 1960s, Murray Gell-Mann and Yuval Ne'eman independently suggested schemes for classifying this huge array of newly discovered particles. Gell-Mann's classification system arose from the fact that the particles he was discovering weren't fundamental particles at all. Instead, they were made up of smaller constituents, which he named *quarks.*

Gell-Mann suggested three types of quarks – up, down, and strange quarks – with certain intrinsic properties. (For some reason, Gell-Mann referred to these three different types of quark, as *flavours* – yum!)

The quarks also had antimatter counterparts, known as *antiquarks*, with opposite charges but the same masses and spins.

Encountering composite particles

So how do these quarks fit together to make up the particles that scientists have known and loved for a long time and these new, unusual particles that were being discovered? It turns out that quarks can be put together in several ways:

- **Mesons** are composite particles that contain a quark and an antiquark, not necessarily of the same flavour; for example, a strange and an antidown quark. The pion and kaon, which we mention earlier in this section, are just two of the many mesons that are known to exist.

- **Baryons** combine three quarks and/or antiquarks and include some of the more familiar particles.
  - Protons are made of two up quarks and a down quark, which gives them their familiar positive charge.
  - Neutrons are comprised of two down quarks and an up quark, giving a neutral particle overall.

Of course, more exotic composite particles exist, and many are based on the baryon recipe, including the delta, omega, and lambda baryons. The omega minus baryon, for example, is formed from three strange quarks. This particle had not been observed when Gell-Mann predicted its existence in 1962, but its subsequent discovery in 1964 cemented the idea of quarks at the heart of particle physics.

Adding more quarks: Charm, top, and bottom

The quark story doesn't end with up, down, and strange quarks. Oh no! Particle physicists soon ran out of possible quark combinations but continued discovering more and more particles.

In 1970, Seldon Glashow, John Iliopoulos, and Luciano Maiani predicted a new quark, known as the *charm quark*. It was observed five years later at the Brookhaven National Laboratory in the USA in the decay of a particle that came to be known as the charmed sigma.

In 1973, Kobayashi and Maskawa predicted two more quarks after considering the decay of a particle called the neutral kaon, which violated the rules as then understood. In 1977, they were proven correct with the discovery of the *bottom quark*. Nineteen years later, its counterpart, the *top quark*, was seen. (Interestingly, someone tried to name the new quarks beauty and truth, but the more boring names top and bottom stuck.)

As with the up, down, and strange quarks, the three new quarks have corresponding antiquarks. This means that the total number of possible quark combinations is immense, leading to the huge number of particles that particle physicists have observed. See Table 9-1 for a summary of known quark types.

According to the Particle Data Group (an organisation that collates information from around the world on the properties of all the particles discovered to date), around 300 different particles have been discovered at present. This number may soar if more exotic quark combinations, such as pentaquarks with five quarks rather than two or three, prove possible.

The ghost particle: The neutrino

Beta rays have long held a place of affection in the hearts of sci-fi writers, but they aren't rays or radiation in the same way as light and X-rays. In fact, beta 'rays' are nothing more than electrons produced when certain elements undergo radioactive decay, such as the decay of caesium-137 into barium-137.

However, the discovery that beta rays are actually electrons gave rise to a problem: The laws of conservation of energy and angular momentum said elements shouldn't decay in this way.

In the 1930s, Wolfgang Pauli and Enrico Fermi argued that another unseen particle is emitted during beta decay, which carries away the rest of the energy and the angular momentum. This particle was eventually christened the *neutrino,* because of its electrically neutral state.

The neutrino's neutrality (which means its tracks can't be directly observed), combined with its minuscule mass and unwillingness to interact frequently with other particles, make the neutrino spectacularly difficult to observe. (In 1956, scientists Fred Reines and Clyde Cowan finally detected neutrinos – actually anti-neutrinos – when they built a detector close to a nuclear reactor in Los Alamos, New Mexico.) Billions of neutrinos produced by nuclear reactions in the Sun pass through your body every second and you don't even notice. In fact, they pass through the Earth with barely a flinch – hence the reason the neutrino is sometimes termed the 'ghost particle'.

Fitting electrons and more into the Standard Model

So where do electrons fit into the quark picture? The answer is they don't. Electrons are thought to be fundamental particles, not made from quarks.

Electrons are part of another family of fundamental particles known as the *leptons.* Like the quarks, leptons come in different flavours, including the electron that scientists know and love, a particle called the *muon*, and another called the *tau* particle, each heavier than the last. Like the quarks, each has its corresponding anti-particle. Table 9-1 lists various leptons that are part of the Standard Model.

Grouping in generation

With the array of quarks and the leptons, scientists finally have a basic toolkit, the Standard Model, to describe the composition of everything in the universe. Table 9-1 arranges all these particles into groups with similar characteristics, known as *generations.*

Table 9-1 The Fundamental Particles of the Standard Model

First generation	Up quark	Down quark	Electron	Electron neutrino
Second generation	Charm quark	Strange quark	Muon	Muon neutrino
Third generation	Top quark	Bottom quark	Tau particle	Tau neutrino

Can you expect a next generation, a fourth group of quarks and leptons? The particle physicists say no. Scientists have carried out various experiments, including a study of the abundance of helium in the universe, which indicates that the number of generations is three and three only. Of course, that hasn't stopped some scientists from looking for new quarks and leptons.

Yet the Standard Model doesn't predict everything. For example, why does more matter than antimatter seem to exist in the universe? No one is really sure. And why do the quarks have the masses they do? Is something happening that scientists have missed?

The Standard Model outlines the basic building blocks of every atom in the universe, but that's only half the story. You also need to know how these particles react to the fundamental forces of nature to really know how the universe came to look as it does today. That's what we look at in Chapter 10.

Chapter 10

Forcing the Pace: The Roles of Natural Forces in the Universe

In This Chapter

- Getting to know the four natural forces
- Identifying the strange particles that transmit forces
- Seeking a theory that unites all the forces of nature

One useful way to think of the universe and its development is as a great play, perhaps one written by Shakespeare.

The play's varied cast of characters is similar to the amazing range of elementary particles existing in the vast universe (check out Chapter 9 for more on elementary particles). But what about the play's dialogue and stage directions that cause the characters to speak and move? In the universe, the fundamental forces of nature perform these tasks, guiding the interaction of elementary particles.

In terms of how much scientists currently know and understand about the universe, humans are somewhere in the middle of the play. Through experiments in massive, expensive particle accelerators, scientists can combine and split apart subatomic particles, essentially winding back the clock of universal time and travelling back towards the play's beginning. Some day, although probably not in the very near future, humans may well discover the first line of the play.

This chapter examines the four fundamental forces that exist in every corner of the universe and impact everything from the tiniest subatomic particle to planets, stars, and even entire galaxies.

Forcing the Issue

What do scientists mean by a *force*? Some forces are easy to understand, such as those including the amount of 'push' you use to get something to move or the strength you have to exert to crack a nut. When you push a child on a swing, for example, you're exerting a force on the swing. Before you begin pushing, the swing is motionless, but when you push, the swing and the child both start moving.

At first glance you may think that any number of different types of force exist. Yet because scientists love categorising things, you can classify every force in the universe as one of four fundamental types: gravity, electromagnetic, strong, and weak.

You've probably heard of the first two forces, but the second two may be new to you. (Without them, however, you wouldn't be here at all.)

- **The strength of the strong force** stops all the atoms in the universe from flying apart. For more on the strong force, see the later section 'Holding things together: The strong force'.
- **The weakness of the weak force** enables stars to burn as slowly as they do, which, for example, has given human life enough time on the Earth to evolve to where things are now without the Sun running out of steam. The later section, 'Venturing deep inside the atom: The weak force', offers more information about the weak force.

Considering Newton's idea of forces

As we discuss in Chapter 3, Isaac Newton's famous laws of motion talk about force. One of the earliest equations you encounter in school science is $F = ma$, which is a shorthand way of saying that force is equal to the mass times the acceleration. (This relationship is known as Newton's second law of motion.)

Although you can easily imagine simple scientific experiments in which you push objects with different amounts of force to see what happens to them, most of the forces in the universe don't act in such a direct, hands-on way.

Gravity, which we talk about in Chapter 3, is one force that exerts itself from a distance. For example, the Sun exerts a gravitational force on the Earth (and vice versa) which keeps the Earth in orbit around its sun, and yet no physical contact exists between the two bodies.

Another example of a force acting at a distance is a magnet and a metallic object. The metallic object is attracted to the magnet and starts moving towards it, with no physical contact existing between the two. What is happening is that the magnet is exerting an electromagnetic force on the object.

Uniting electricity and magnetism: The electromagnetic force

In Chapter 4 we describe how James Clerk Maxwell realised that electricity and magnetism are closely related and are, in fact, just two manifestations of the same thing – *electromagnetism.*

Like gravity, the electromagnetic force between two electrically charged particles obeys an *inverse square law*. This law means that two electrically charged particles, 1 metre apart, experience a force four times stronger than if they are 2 metres apart (because $4 = 2 \times 2$) and nine times stronger than if they are 3 metres apart (because $9 = 3 \times 3$).

When people say that opposites attract, they're most likely talking about human relationships. But the concept of opposites attracting probably came from the world of electromagnetism where negative charges are attracted to positive ones and north magnetic poles are attracted to south ones. Electrical charges and magnetic poles of the same type repel each other because negatively charged electrons tend to whiz away from one another. Try to put the red ends of two magnets together and you see what we mean.

Seeing the electromagnetic force everywhere

You may be surprised to hear that electromagnetism is behind many of the forces you know well in everyday life.

Imagine you want to push something – a child on a swing, say. Both you and the swing are made up of atoms. As we describe in Chapter 9, an atom is just a central positively charged nucleus and a cloud of negatively charged electrons around it, making the atom electrically neutral overall.

But as the atom at the very extremity of your finger approaches the closest atom on the swing, something odd happens. At some point, the electrons in the atom on your finger are closer to the electrons in the outer atom of the swing than they are to the protons in the atomic nucleus of the atom of your finger. Because the electrons of your finger and the swing are both negatively charged, they start to repel each other. Because you are standing still, the atoms in the swing get pushed away and the child giggles with delight.

The electromagnetic force is also behind *friction*, the resistance that results from one object being in contact with another (with reasoning similar to the preceding child-in-a-swing example) and chemical reactions (exchanging or sharing of electrons between atoms is essential to chemistry).

Virtual particles

Light and other electromagnetic radiation can act like a wave on the one hand and as a stream of particles, known as *photons*, on the other (see Chapter 9). Scientists say that the electromagnetic force is *mediated* by the photon, meaning that the photon carries the force.

Many of the particles that mediate forces are virtual particles, which work in a very curious way that you can't directly observe. These virtual particles come about by 'borrowing' energy, seemingly from nothing.

To think about this borrowing process, imagine for a moment that you are a decent human being who decides to collect money for your favourite charity, The Poor Particle Physicists Campaign. You have a good afternoon and manage to collect £100 for those poor quantum physicists who are on their uppers. The area manager for the charity says he's going to come round tomorrow night to collect your takings.

The next morning you hear a knock on the door: Someone is collecting for another charity, The Cosmologists' Society. This charity is one of your pet causes as well and you want to give generously. Then you remember that you don't have any money in the house – or do you? You nip into the bedroom, crack open the collecting tin for The Poor Particle Physicists Campaign, and give £50 in small change to the collector for the cosmologists.

You feel a bit naughty but know that by the time the area manager for The Poor Particle Physicists Campaign comes round in the evening, you can get some more money out of the bank before the manager is any the wiser.

Something similar happens in the world of particle physics – but rather than borrowing money for a time, energy can be borrowed. Mother Nature is perfectly happy for elementary particles to borrow a bit of energy for a while as long as it's paid back before she notices. Not only that, she even lets elementary particles go overdrawn, seemingly without limit. The only restriction is that the more energy that's borrowed, the less time it can be borrowed for.

Based on Einstein's equation $E = mc^2$, you know that mass and energy are interchangeable. This truth leads to one of the most incredible ideas of particle physics: Because subatomic particles can borrow energy for a short amount of time, they can also create matter in the form of particles for fleeting moments. These particles are known as *virtual particles* and scientists know they exist because they can observe the real particles that they turn into. Beta decay, the production of electrons and anti-neutrinos from the nuclei of atoms, relies on a virtual particle to happen, for example.

So where does this energy come from? It would be easy to assume that the energy is the kinetic energy of any original particles involved in an interaction, but that's not the case. In fact, the energy comes from space itself, even if that space is empty. Conceiving that a vacuum can lend energy is hard, but that's exactly what happens. In fact, many cosmologists believe that energy fluctuations in the vacuum may have kicked off the Big Bang process in the first place and provided the variations that we now see as small differences in the temperature of the cosmic microwave background radiation (see Chapter 6).

Where is all this leading? Well, virtual particles are a handy way of thinking about all four of the fundamental forces.

As we discuss in Chapter 9, Einstein came up with the idea that electromagnetic radiation can be *quantised* – carried in convenient packets of energy that can then eject electrons from the surface of certain types of metal if those packets are energetic enough. The idea of electromagnetic radiation being a particle can also explain how all the fundamental forces operate.

Consider a typical interaction between some particles that involves electromagnetism. *Electron scattering* is the process by which electrons are nudged off their original paths through space by their interactions with other charged particles. Figure 10-1 uses the collision between some snooker balls to illustrate the process of an electron scattering off another electron.

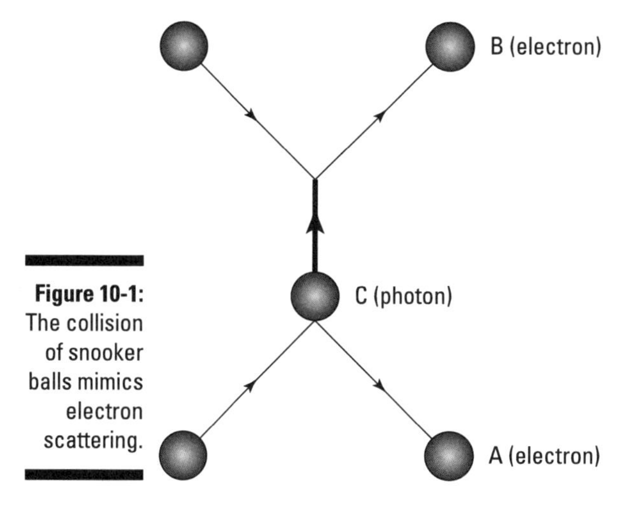

Figure 10-1: The collision of snooker balls mimics electron scattering.

Worth the price tag?

When confronted with the idea of particle accelerators, many people – particularly politicians and taxpayers – ask, 'What's the point?' Accelerators are undeniably expensive things. When the Large Hadron Collider (due to start up at the CERN laboratory in Geneva in 2008) was first given the go-ahead in the early 1990s, the expected cost was around £1.2 billion and has risen since then.

The reason for spending money on expensive particle accelerators is nothing less than understanding the very nature of the universe. Scientists believe that at high energies the fundamental forces are unified, and by using the high energies reached by smashing particles together in accelerators, they hope to be proven right.

But proving the theory isn't enough on its own. What particle physicists want to achieve with expensive machines is to recreate the universe's earliest moments, albeit for a very short time. They hope that by achieving these high energies, they can see why the universe turned out as it has and, even more fundamentally, how the universe came into being in the first place.

In the figure, two balls (A and B) are moving across a surface. Ball A hits a third ball C, which then follows the path of the thick middle line and hits incoming ball B. Balls A and B are both diverted onto different paths as a result of the collision.

Now imagine that the figure represents an interaction between electrons. Balls A and B now represent electrons while ball C represents a photon that carries energy between the two electrons. Scientists believe that the fundamental forces can be considered to operate in just this fashion.

In the electron scattering interaction represented by Figure 10-1, the photon involved is a virtual particle. You couldn't observe the photon with, say, a camera. However, you can see that the two electrons have been diverted off their original paths, proving the existence of the virtual particle at the heart of the interaction.

Richard Feynman's squiggly diagrams

Richard Feynman, born in 1918 in New York, was a showman as well as a scientist. Famously, during the inquest into the Challenger shuttle disaster, he dropped a rubber ring into a cup of icy water to show how a rubber O-ring seal in Challenger's fuel tank may have fractured.

Feynman was also one of the key figures in the development of particle physics as a discipline in its own right. Among his most brilliant ideas is a way of visualising interactions between elementary particles, similar to the snooker ball interaction in Figure 10-1.

In Feynman's diagrams, straight lines represent *fermions* (particles such as the electron, proton, neutron, and the neutrino – see Chapter 9 to find out about all these) and wavy lines represent *bosons* (particles such as the photon and some others which we discuss in the section 'Mediating the weak force'). The exception is the Higgs boson, which is represented by a dotted line (see the later section 'Giving Things Mass: The Higgs Field and Boson') and particles called *gluons,* which are represented by looped lines.

Feynman's diagrams feature two axes: Left to right shows how an interaction develops in time, whereas bottom to top depicts how the interaction develops in terms of space, or position. Figure 10-2 shows a simple Feynman diagram representing two electrons interacting with each other.

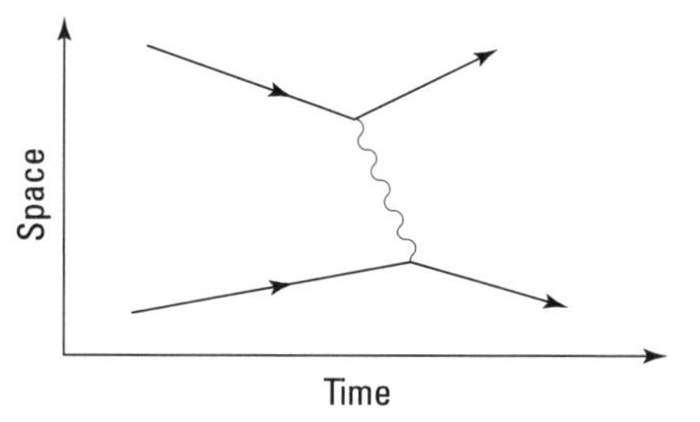

Figure 10-2: A simple Feynman diagram.

If you run your finger from left to right, you can follow the electrons' interaction through time and see the two electrons coming closer together, exchanging a photon (indicated by the wiggly line), and then moving off in a different direction.

Feynman diagrams are used by particle physicists to calculate how likely (or how frequently) certain processes will happen (for example, the decay of a muon into an electron, a neutrino, and an anti-neutrino).

How can we do this with these sketches with wiggly lines? The idea is that you have to draw every possible Feynman diagram that has the correct particles going in and coming out (in the case of the process mentioned above a muon, an electron, a neutrino, and an anti-neutrino).

Doing so gets complicated because all sorts of processes may happen in the course of an interaction, including the creation of pairs of virtual particles and their subsequent annihilation before anyone can notice, as Figure 10-3 shows. These virtual pairs are made up of a particle and its antiparticle, which has the same mass but many other characteristics, such as electric charge, opposite to its particle counterpart – for example, an electron and a positron.

In a spin: Fermions and bosons

As well as giving them odd names, physicists also classify particles into two categories depending on their spin. *Spin* is a measure of the amount of angular momentum a particle has (*angular momentum* is a measure of the ease or difficulty of slowing down or speeding up something that's rotating).

Like some other properties of subatomic particles, their angular momentum is *quantised* – that is, can only take certain fixed values – according to the following equation:

$$h\sqrt{\frac{s(s+1)}{2\pi}}$$

where s is the so-called spin quantum number and h is Planck's constant. For every particle ever observed, s can only be an integer (0, 1, 2, 3, and so on) or a half-integer (½, 3⁄2, 5⁄2 and so on).

- **Bosons,** such as photons and W and Z particles, have integer spin and their behaviour is governed by something called Bose-Einstein statistics.
- **Fermions,** such as electrons and neutrons, have half-integer spins and their behaviour is governed by something called Fermi-Dirac statistics.

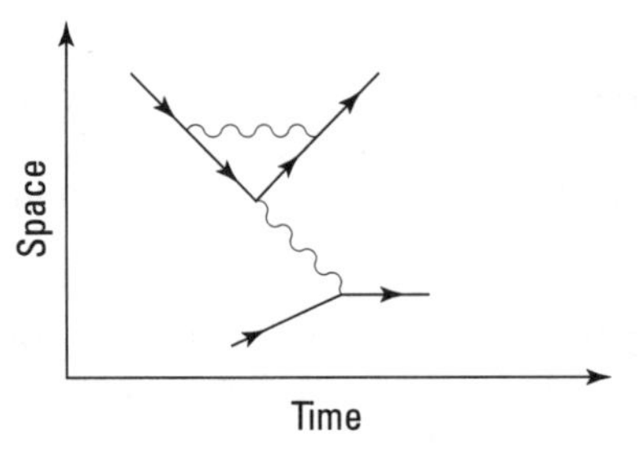

Figure 10-3: A Feynman diagram showing the creation of a pair of virtual particles.

Figure 10-3 is a Feynman diagram of the same electron scattering event that appears in Figure 10-2. The lines coming out of the diagram – that is, the real particles you can observe – are the same.

Feynman's big contribution was realising that every line and *vertex* (points where lines meet) in these diagrams corresponds to a fixed number when translated into a normally extremely complicated mathematical equation. Previously, these equations had been almost impossible to solve but Feynman's diagrams provided a simple way of doing so.

Venturing deep inside the atom: The weak force

The history of the weak force begins with the discovery of radioactivity by Henri Becquerel in the last years of the 19th century and, in particular, the discovery of uranic rays, which subsequently became known as *beta decay* or *beta radiation*.

Scientists now know that beta decay is a stream of energetic electrons, which have been expelled from the atomic nucleus. But at the time of Becquerel's work, the known forces of nature didn't explain what was happening.

Over the next three decades, scientists began to realise that another fundamental force – and a very weak one at that – is at work deep inside the atom. In 1930, Wolfgang Pauli speculated that a new particle, known as the neutrino, is involved in the process of beta decay in order for the process to conform to the laws of conservation of energy and momentum (see Chapter 9 for more details).

In 1934, the Italian physicist Enrico Fermi submitted a paper to the journal *Nature* explaining the theory behind beta decay. Fermi's idea is that a process is going on in the middle of a nucleus that's similar to an excited electron emitting an electromagnetic photon. According to Fermi, the proton emits some force-carrying particle similar to the photon, which causes the neutron to turn into a combination of a proton, an electron, and an anti-neutrino.

The rates at which beta decay occurs showed Fermi that if his hypotheses were correct, the force is very weak compared to the strong force that normally keeps the nucleus together (which is how the weak force got its name, appropriately enough).

Holding things together: The strong force

In the earlier section, 'Uniting electricity and magnetism: The electromagnetic force', we talk about the electromagnetic force and how similarly charged particles repel each other. 'Hang on a minute,' you cry, 'What about the atomic nucleus itself?'

Okay, maybe you didn't, but think about how the nucleus is constructed (check out Chapter 9 for more on the nucleus). Rutherford's model of the atom includes a nucleus made up of protons and neutrons. *Protons* are positively charged, and *neutrons* are, well, neutral. This arrangement means that if electromagnetism were the only force involved in creating an atomic nucleus, the nucleus would fly apart because the positive charges within the nucleus would repel each other.

But this doesn't happen because another, stronger, force – named, as you've no doubt guessed, *the strong force* – keeps the nucleus together. If the strong force didn't exist, atoms wouldn't have stuck around for very long. (Actually a deeper process is going on within the nucleus involving quarks, and we discuss this process in the later section, 'Mediating the strong force'.)

Uniting the Forces of Nature

At first consideration, the electromagnetic force and the weak force (also called the *weak interaction*) may seem to be totally unrelated to each other.

However, in the 1960s three physicists – Sheldon Glashow, Abdus Salam, and Steven Weinberg – came up with the idea that in very high energy situations (soon after the Big Bang, for example), the electromagnetic force and the weak interaction are just two different ways of looking at the same thing, which they dubbed the *electroweak interaction.*

The electroweak interaction isn't as odd as it sounds. After all, who would have thought that electricity and magnetism are the same thing before James Clerk Maxwell came along and proved it? Glashow, Salam, and Weinberg suggested that the weak force is mediated by a force-carrying particle called a *boson,* in the same way that the electromagnetic force is mediated by photons.

Mediating the weak force

However, three different types of boson are required to account for three different types of weak interaction:

- **W^-**, which is a neutral particle that decays into a positively charged one (such as the neutron transforming into a proton through beta decay).
- **W^+**, which is the anti-particle of W^- and allows a positively charged particle to decay into a neutral particle (such as the proton transforming into a neutron by emitting a positron and a neutrino).
- **Z**, which causes the electrical charge to remain the same (such as the scattering of muon neutrinos, the so-called second generation neutrino that we discuss in Chapter 9).

Making things colourful

In the original conception of the existence of quarks (Chapter 9 has more on quarks), Oscar Greenberg of the University of Maryland suggested that quarks come in three colours (red, green, and blue) and three anti-colours (anti-red, anti-green, and anti-blue). Additionally, each of the quarks in the proton and the neutron has one of these colours.

Now, the property of colour has nothing to do with the colours you observe in everyday life. Instead, Greenberg was making an analogy with what happens when you mix red light, blue light, and green light together – you get white (colourless) light. In the same way, combining quarks of various colours essentially cancels out the property of colour.

When scientists talk about the force that acts between quarks of different colour charges, they really should be talking about a colour force rather than a strong force, but the latter name has stuck around and now refers to both the interaction between protons and neutrons and the interaction between quarks.

In 1973, an experiment at the European Particle Physics Laboratory CERN called Gargamelle (named after a giantess in the literary works of Rabelais) found the first piece of evidence to show that Glashow, Salam, and Weinberg are correct. Within a giant vessel filled with super-heated transparent liquid, known as a *bubble chamber*, scientists used the refrigerant freon to detect neutrinos, and they observed muon neutrino scattering. The rate at which the process occurred matched the predictions of Glashow, Salam, and Weinberg, hinting that their ideas are correct.

In 1984, CERN's Super Proton Synchrotron – which was able to smash protons into anti-protons at extremely high energies – observed the production of both W and Z bosons. These bosons had masses very close to what Glashow, Salam, and Weinberg had predicted, confirming their electroweak interaction theory.

W and Z bosons have large masses, as Glashow, Salam, and Weinberg predicted. The large mass of these bosons explains why beta decay happens so much less frequently than similar electromagnetic interactions and also why the weak force weakens much faster than the electromagnetic force. Why? Because W and Z bosons are virtual particles with large mass (or energy), they can survive for only a fleeting moment before Mother Nature wants payback (as we discuss in the previous section 'Virtual particles').

Mediating the strong force

After scientists realised which particles mediate the electromagnetic force and the weak force, they then wanted to know what mediated the strong force.

Instead of giving boring letter names to the particles that mediated the strong force, scientists came up with something that really does what it says on the tin.

The mediator of the strong (colour) force is known as the *gluon*. Gluons change quarks from one flavour to another and from one colour to another (check out the sidebar 'Making things colourful') – and so a red up quark can change into a blue down quark.

Scientists believe that gluons are massless (like the photon) and don't carry electrical charge (like the two W bosons of the weak force). However, gluons do have colour, just like the quarks, which means that gluons experience the strong force as well as mediate it. Because gluons experience the strong force in addition to carrying it, they are extremely difficult to study mathematically, which is one reason why scientists know a lot less about the strong force than the other forces.

Evidence for the existence of gluons was first seen at the Stanford Linear Accelerator (SLAC) in the USA and subsequently at the DESY experiment in Hamburg.

Both gravity and electromagnetic forces diminish as you get farther away from the source of the objects generating these forces. For example, the gravitational and electromagnetic forces between two protons at opposite ends of the universe are so impossibly tiny that they can be ignored. The strong (colour) force between quarks is different – it doesn't get any weaker as the quarks attempt to move apart. What this means in practice is that the quarks can't move outside the bounds of the particle containing them, a phenomenon known as *confinement*. This reason is why you never see a quark on its own.

Considering quantum gravity

With so many bosons flying around the universe and mediating various forces, scientists soon suggested that maybe the force of gravity is transmitted in the same way as the other forces – by a gravity-carrying particle dubbed the *graviton*.

Comparing the fundamental forces

The following table lists all four of the fundamental forces, the particles that carry each force, their relative strengths, and their range.

Force	*Mediator (Force-Carrying Particle)*	*Relative Strength*	*Range (Metres)*
Strong	Gluon	1	10^{-15}
Electro-magnetic	Photon	10^{-3}	Infinite
Weak	W and Z bosons	10^{-5}	10^{-17}
Gravity	Graviton	10^{-38}	Infinite

To give a clear sense of the relative strengths of the forces, the table lists them relative to the strongest force – the strong force. The strengths vary depending on the sort of particles that are interacting, but the listed values are good approximate figures.

Looking at the strength column, you can also see that the weak force is a misnomer. Gravity is by far the weakest force – although it makes up for what it lacks in oomph by being felt across the entire universe.

Although the graviton has never been observed, scientists know that if it exists it must be massless – like the photon – because gravity has an unlimited range of influence.

At present the best hope of finding gravitons is in experiments that try to find evidence of *gravitational waves.* (Scientists assume that gravitons exhibit the same wave-particle dichotomy as photons).

The main problem with trying to pin down the graviton is that gravity is incredibly weak compared with the other three fundamental forces. Some time may well pass before any evidence of gravitons can be uncovered.

Giving Things Mass: The Higgs Field and Boson

One of the big puzzles of the Standard Model (we talk about the Standard Model in Chapter 9) is why things have the amount of mass they do. For example, why is the top quark many thousands of times heavier than the electron?

Scientists don't like these extreme variations in particle masses. Such differences make scientists wonder whether they really have reached the most fundamental level possible in terms of explaining the universe.

One scientist who tried to come up with an explanation was Peter Higgs, in the 1960s, who theorised why the W and Z bosons (see the previous section 'Mediating the weak force') are so massive and the photon (which electroweak interaction theory says is closely inter-related to the W and Z boson) is massless.

Higgs proposed something called *the Higgs field*, which permeates all the universe. The mass of any particle, such as the electron, depends on how strongly it interacts with this shadowy – and as yet unobserved – field. The interaction that takes place involves the exchange of yet another boson, *the Higgs boson*, which is also unobserved.

You may be thinking that this notion is ridiculous, but stick with us for a moment more. To get an idea of what the Higgs boson means in practice, consider the 1993 challenge presented by Britain's minister for science, William Waldegrave. Waldegrave asked scientists to answer – on a single sheet of paper – the question, 'What is the Higgs boson, and why do we want to find it?' Waldegrave promised the winning entries a bottle of champagne.

One of the winners was Professor David Miller of University College London, who envisaged the Higgs field as a party attended by then Prime Minister Margaret Thatcher:

> *Imagine a cocktail party of political party workers who are uniformly distributed across the floor, all talking to their nearest neighbours. The Prime Minister enters and crosses the room. All the workers in her neighbourhood are strongly attracted to her and cluster round her. As she moves she attracts the people she comes close to, while the ones she has left return to their even spacing. Because of the knot of people always clustered around her she acquires a greater mass than normal; that is, she has more momentum for the same speed of movement across the room. Once moving she is hard to stop, and once stopped she is harder to get moving again because the clustering process has to be restarted.*

Miller went on to suggest that the Higgs boson was similar to a rumour spreading across the same room, with the clustering of people also giving the Higgs boson mass.

Most particle physicists believe that the Higgs field and boson do exist and are sure to be found sooner rather than later. Indirect hints have popped up in existing experiments, suggesting that scientists may find the Higgs boson (sometimes called 'the God particle' by sceptics) in the next generation of particle accelerators, such as the soon to be operational Large Hadron Collider (LHC). When LHC is switched on, the world may know for sure.

Searching for GUTs and TOEs: Grand Unified Theories and Theories of Everything

Some scientists now believe that the four fundamental forces known today were at some point in the past just different manifestations of a single, unified force. Einstein himself believed in the existence of some *grand unified theory* (or GUT) that explained everything. He spent the last two decades of his life trying to discover it.

Theories of everything (or TOEs) go one step farther than GUTs and hope to explain in one overarching theory all physical processes, as well as why certain physical properties have the values they do. Sadly, humans are still a long way from such theories, and little evidence exists to suggest that the idea of TOEs is correct in any case.

One of the big problems that all GUTs and TOEs face is the difficulty in making quantum mechanics square up with Einstein's general theory of relativity. Thus far, most attempts to make quantum mechanics and general relativity compatible have failed. Those theories that manage to do so have had to employ weird and wonderful notions (even weirder and more wonderful than quantum mechanics) to explain things and have failed to produce any concrete tests that prove them.

How much energy do scientists need?

According to the electroweak interaction theory (see the earlier section 'Uniting the Forces of Nature'), the electromagnetic and weak forces are one and the same at extremely high energies. The success of the electroweak theory makes scientists hopeful that they can unify the other forces in a similar manner.

In 1974, Glashow and Howard Georgi suggested that the electroweak and strong forces may be different aspects of the same thing. However, proving this hypothesis is difficult because scientists don't currently have enough energy.

According to the latest guesses, the energies required to unify the strong and electroweak interactions are enormous – something like 10^{16} GeV. Currently, the highest energy that CERN's LHC particle accelerator can achieve is approximately 10^{6} GeV. Adding quantum gravity to the pot is going to require even higher energies – some 10^{19} GeV according to some scientists. You can see that scientists aren't going to reach those energies for some time to come.

By pushing particle accelerators to still higher energies, scientists are hoping to reunite the four fundamental forces, just as these forces were at the beginning of the universe. By examining the aftermath of a collision of two high energy particles at these extremely high *unification energies,* scientists one day hope to see what happened in the first few crucial fractions of a second after the universe began.

Chapter 11

Shedding Light on Dark Matter and Pinging Strings

In This Chapter

- Looking beyond the directly observable
- Exploring dark matter and energy
- Going beyond four dimensions with string theories

Astronomy has its roots in the science of optics. The very first astronomers were scientists interested in the optical properties of light – that is, how light is reflected by mirrors, focused while passing through lenses, and bent and split as it passes through prisms.

Even in the centuries after the invention of the telescope, astronomers continued to focus on light. But with the realisation in the 1800s that light was just one type of electromagnetic radiation (see Chapter 4), astronomers started wondering what the skies looked like by means of instruments other than optical ones such as the eye and the telescope.

As a result, much modern knowledge of the stars comes from instruments that can't 'see' in the everyday understanding of visible light. These instruments include the radio telescope and X-ray detectors, for example.

Yet beyond what you can detect through the various kinds of electromagnetic radiation, a whole other universe appears to exist – one made up of hidden and mysterious things, such as dark matter, dark energy, and minuscule strings that vibrate away in spatial dimensions invisible to us in everyday life.

In this chapter, we shed light on some of these mysteries – and in doing so, find out more about how the universe came into existence.

Addressing the Dark Elephant in the Room: Dark Matter

In the last few decades, cosmologists have become increasingly aware of an elephant in the room. Not a real one, mind you, but a big scientific problem that they've been ignoring for many years but can't avoid any longer.

The problem is matter – or more precisely, the lack of it. For all the wonderful theories about how the Big Bang happened, cosmologists are beginning to realise that the matter that you can see – in the form of stars, galaxies, and clouds of dust and gas – just isn't enough to explain how the universe came to look like it does today given its supposed age.

Scientists have come to believe that what you see – or at least what you see directly – makes up just a small proportion of everything that exists in the universe. The universe, it seems, has a very dark secret – one that is only now coming to light.

Discovering the dark side

In Chapter 8, we discuss the Swiss astrophysicist Fritz Zwicky and his tired light theory, which set out to explain away the observed red-shifts of distant galaxies. Although Zwicky was wrong about the reasons for red-shifts, he may have been right about other ideas.

In 1933, Zwicky was looking at the Coma cluster of galaxies using the 45-centimetre (18-inch) Schmidt telescope at the Mount Palomar observatory in the United States. He measured the *radial velocities* – the velocities in the line of sight – of eight galaxies in the cluster and found an astonishingly wide range. In fact the difference between the fastest and the slowest velocities was something like 1,000 kilometres (620 miles) per second.

Zwicky used a piece of maths called the *virial theorem*, which shows how the motion of objects in a system is related to the forces acting upon them – in this case, how galaxies move under the force of their mutual gravitational attraction.

Zwicky used the virial theorem to work out the average mass of a galaxy in the cluster and came up with 4.5×10^{10} times the mass of the Sun. Yet the average luminosity of a galaxy in the cluster was only 8.5×10^{7} times that of the Sun, meaning that the ratio of luminosity to mass was something like 500 times higher than the same ratio for the average star in our local area of the universe (calculated by observations of their light and motion due to gravity). This was much higher than expected and left Zwicky – who assumed that these ratios should be similar – with a puzzle to unravel.

Speculating on the reasons for this odd finding, Zwicky suggested that more matter must exist in the Coma cluster of galaxies than the total luminosity suggested. Due to the Coma cluster's observed gravitational effects, some other *dark matter* that wasn't directly observable must exist. See the later section 'Defining dark matter' for more on this subject.

Noting strange galaxy rotations

The real interest in dark matter began in the late 1970s when astronomers began noticing something odd about the way that stars rotate around the centres of spiral galaxies.

In spiral galaxies, most stars are arranged in a kind of flattened circular disc that spreads out from a central bulge, or nucleus, a little like the shape a pizza base makes when a show-off cook flings it spinning into the air (the colour section has a photograph of a spiral galaxy and Chapter 13 has a diagram).

Based on Newton's universal law of gravitation (check out Chapter 3 for all about Newton), stars in the circular discs of spiral galaxies should move more slowly the farther out they are from the centre (because the force due to gravity follows an inverse square law). However, observations of spiral galaxies show that this isn't the case. Outside the central core of the galaxy (its *nucleus*), the velocities of stars are remarkably constant.

The most plausible explanation for the consistency in velocities is that a large amount of unseen matter – that is, dark matter – lies far from the galactic nucleus, affecting the rotational speed of the galaxies.

Dark matters

In Chapter 7, we talk about the critical density of mass in the universe. The density is called critical because it determines whether you're living in an open, closed, or flat universe.

In an open universe, matter is insufficient, by the force of gravity alone, to stop the expansion caused by the Big Bang. The opposite is the case for a closed universe. A flat universe sits between these two extremes – the expansion is eventually going to stop but only after an infinite amount of time.

Most of the evidence points to the universe being flat, but the amount of dark matter that exists in the universe is key to knowing for sure which of the three types of universe humans live in.

Defining dark matter

We talk in the preceding section about the rationale for dark matter but what, exactly, is it? Broadly, two leading ideas persist:

- Dark matter is a new form of matter that doesn't emit or absorb light and other forms of electromagnetic radiation. Scientists refer to these particles as WIMPs, for reasons we explain in a moment.
- Dark matter is ordinary matter that's so dim humans don't see it. Scientists refer to these objects as MACHOs. (Who said scientists don't have a sense of humour?)

What a WIMP!

The first suggestion to explain dark matter is based on WIMPs, or *weakly interacting massive particles*.

We discuss the fundamental forces of nature – weak, strong, electromagnetic, and gravity – in Chapter 10. One suggested explanation for dark matter is that an as yet undiscovered class of particles exists that only experience the weak force and gravity. The inability of these WIMPs to interact through the electromagnetic force explains why humans can't see them.

Scientists also predict that this form of dark matter is very heavy, perhaps 10 to 100 times the mass of the proton – very high for a fundamental particle. This huge mass means that WIMPs are sluggish, making them into potential 'seeds' on which galaxies can grow, and could also account for the discrepancy in the mass-luminosity ratio observed by Zwicky.

A considerable number of experiments are currently in progress in an attempt to detect WIMPs.

- Some scientists are trying to observe WIMPs directly, by spotting the vibrations they trigger as they bounce off the nuclei of exotic metals such as germanium in Earth-based detectors.
- Other scientists are looking for WIMPs indirectly. One theory suggests that stars may contain large numbers of WIMPs and anti-WIMPs (that is, their anti-particle counterparts). In theory, WIMP-anti-WIMP annihilations in the heart of these stars send out a stream of particles, particularly neutrinos, which may be detectable on the Earth.

So far, results have been inconclusive with both lines of attack.

Acting MACHO

Perhaps all this talk of weird and wonderful new particles is a bit too much to take. Some scientists think that the idea of WIMPs is just tommy-rot and that much more conventional rationales can explain the missing matter in the universe.

In this theory, MACHOs – or *massive compact halo objects* – are celestial objects that just happen to be dark and, as a result, are hard to detect. Candidates for MACHOs include burned-out stars, brown dwarfs (pseudo-stars that don't have enough mass to ignite nuclear fusion), planets that have been knocked out of their solar systems, and just about any other sort of space junk that doesn't emit much light or other radiation, including mini black holes.

As with WIMPs, MACHOs have never been observed, despite countless experiments to verify their existence. The problem in observing MACHOs is the same as for WIMPs – MACHOs don't emit much light or other radiation.

Several experiments use a technique known as *gravitational microlensing* to spot MACHOs. Gravitational lensing (which was suggested by Einstein, as we discuss in Chapter 4) is where the gravity of a massive object distorts light from distant stars and galaxies that lie beyond it, sometimes enabling you to see objects that are otherwise hidden. *Gravitational microlensing* is this same phenomenon, only on a smaller scale. Some scientists believe that they can observe MACHOs by looking for lensing of very bright distant objects.

Researchers have identified MACHOs in the halos of galaxies using microlensing, but whether enough of them are out there to account for the effects of dark matter is questionable.

Mapping the dark matter of the universe

Evidence for dark matter is growing daily and some of the clearest signs of its existence came in 2006 when researchers at the Chandra X-ray observatory looked at the 'bullet cluster' – two clusters of galaxies colliding.

Looking at the collision, the scientists could see that the stars of the galaxies weren't dramatically affected, mostly passing through the collision point, slowed only by gravity. Hot gases that existed between the galaxies, on the other hand, interacted more with each other, slowing down more substantially.

Using gravitational lensing, the researchers could see that dark matter in the galaxies wasn't slowed at all. The lensing was strongest in two separated areas near the galaxies. A photograph in the colour section shows the distribution of dark matter in the universe.

Getting Even Darker: Dark Energy

As if one dark and mysterious thing in the universe – the dark matter we discuss in the preceding section – isn't enough, at the end of the 1990s astronomers found that they need another equally mysterious thing in order to explain the universe.

Observations of very distant supernovae by scientists on the Supernova Cosmology Project, led by Saul Perlmutter at the Lawrence Berkeley National Laboratory, and the High-z Supernova Search Team, led by Brian Schmidt at the Australian National University, found that these supernovae are dimmer than scientists expected.

Supernovae are the enormous explosions that signal the deaths of stars. They come in different types, depending on how they occur. The ones observed by the two teams are called type-Ia supernovae. These explosions occur to *white dwarfs*, small compact stars that are left over after a star has finished burning hydrogen, bloated out to become a red giant, and then shed its outer layers as a planetary nebula. See Chapter 12 for more on this type of supernova.

Normally such stars continue merrily along without doing much for the rest of their days. However, if the white dwarf is part of a two-star system, it can start stealing material from the other star. If this extra material takes the star over about 1.4 times the mass of the Earth's Sun – the so-called *Chandrasekhar limit* – the star can undergo a spectacular gravitational collapse that triggers the supernova explosion.

Astronomers like this type of supernova because they can be used as standard candles, much like the Cepheid variables we describe in Chapter 5. That is, scientists can use type-Ia supernovae to measure cosmic distances.

The teams realised that the galaxies in which type-Ia supernovae were happening were much farther away than expected, based on the Hubble expansion of the universe. The only explanation the teams came up with was that the expansion of the universe is accelerating. This fact is very troubling indeed, because in a universe comprised of matter alone expansion should slow down, not speed up.

The astronomers realised that whatever was causing the acceleration wasn't something visible. Measurements of the cosmic microwave background by NASA's Wilkinson Microwave Anisotropy Probe satellite have since shown that this force isn't in the form of matter but energy – *dark energy*.

The concordance model

If the cosmological constant theory is correct, the theory of how ordinary matter, dark matter, and dark energy are mixed up in the universe is known as the concordance, or Lambda-CDM, model. The latter name is a bit of a mouthful but comes from a combination of the mathematical symbol for the cosmological constant (lambda, λ) and the abbreviation for cold, dark matter (CDM).

The current concordance model seeks to explain how the universe came to look like it does today on the basis of various proportions of ordinary matter, dark matter, and dark energy in the universe. These proportions are given in terms of *energy density*, or how much energy is present in a given volume of space. General relativity says that the energy density and the expansion of the universe are intimately connected. Scientists currently believe that ordinary matter contributes just 4 per cent of the energy density of the universe, dark matter 22 per cent, and dark energy the remaining 74 per cent.

The worrying thing here is that 96 per cent of everything in the universe is stuff that scientists know almost nothing about.

Like many of the concepts in cosmology, dark energy is hypothetical. It hasn't been directly measured, but models based on assuming that it does exist fit reasonably well with other observations of the universe.

Very little is known about dark energy other than the fact that it must be capable of causing a weird gravitational effect – even on the enormous scale of the distances between galaxies – that makes things move away from each other, rather than closer together. This effect is called *gravitational repulsion*.

The two leading explanations for dark energy are the cosmological constant and quintessence.

Cosmological constant

As we discuss in Chapter 6, Einstein introduced the idea of the cosmological constant into his equations for general relativity to explain away the expansion of the universe. He later said that doing so was his greatest blunder. However, if the existence of dark energy is proven beyond doubt, Einstein may yet have the last laugh.

In real terms, the *cosmological constant* means that the vacuum of space is not as empty as you may imagine and that even with nothing there, it possesses some basic energy of its own.

If the vacuum does have its own energy, this energy can explain the accelerated expansion of the universe. Non-zero *vacuum energy* leads to the possibility that matter can be created out of 'nothing' because energy and mass are interchangeable – although they would have to be in particle-antiparticle pairs because of various conservation laws.

Some cosmologists believe that a link exists between this vacuum energy and the *Higgs particle*, the Standard Model particle responsible for giving objects mass (refer to Chapter 10 for an explanation of this scary sounding sentence).

Quintessence

Many scientists see problems with the cosmological constant and the concordance model. For example, if a constant vacuum energy exists, the cosmological constant should be very large. Yet observations seem to suggest that it's very small. The difference between the theoretical prediction and the real-life observations is enormous – of the order of 10^{120} – and so a rather inconvenient problem exists.

To try to address the apparent difference, some scientists have come up with the idea that the cosmological constant isn't constant at all, but has changed through time, leading to different levels of vacuum energy in different parts of the universe. This concept is called *quintessence* and is a less popular alternative to the concordance model.

Obviously, science is still far from knowing which explanation is correct.

Stringing the Universe Along

If the puzzles of dark matter and dark energy seem crazy, hold on to your hats. Searching for deeper understanding of the way the universe operates has led many scientists on even wilder theoretical adventures.

We talk in Chapter 10 about *theories of everything*, which unite all the forces of nature. One of the biggest problems facing scientists trying to create such theories is the seemingly unbridgeable gulf between quantum theory and general relativity.

However, one contender for such a far-reaching theory has a larger number of supporters than any other – string theory.

Measuring a piece of string

As far as scientists can ascertain, quarks and leptons (see Chapter 9) are truly *fundamental* – they don't seem to have any inner structure. Yet string theorists argue that the internal structure of quarks and leptons are too small to be perceived with the current limits on technology.

The big idea behind string theory is that interactions between fundamental particles don't just happen at precise points in space but are spread out over small areas. The Standard Model (see Chapter 9) says that fundamental particles are point-like objects – that is, they have zero dimensions as opposed to the everyday three dimensions of space or four dimensions of space and time. In fact, string theorists say that particles are really *line-like* (one-dimensional objects), which means that they can vibrate at certain fixed frequencies, rather like the string on a musical instrument such as a guitar.

Figure 11-1 shows the Standard Model interaction at a point and a string theory interaction. Notice how the string interaction is smeared out over spacetime.

String theorists believe that their assertions can bridge the seeming incompatibility of general relativity and quantum mechanics. Whereas the equations of general relativity 'blow up' at zero distances, the equations remain sensible on the very short size scales of strings.

The other attractive feature of string theory is that different vibration frequencies of the strings represent different types of fundamental particle. Particle physicists have always been somewhat uncomfortable with the large number of fundamental objects in the Standard Model. String theory reduces the number of fundamental objects to one.

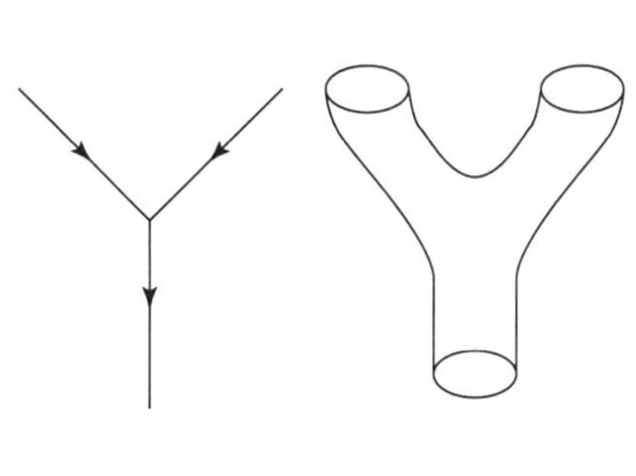

Figure 11-1: How string and particle interactions differ. A Standard Model interaction at a point (left) against a string theory interaction (right).

Comparing competing string theories

At this point, string theory may sound nice and simple, but like a tangled ball of string, things can easily get complicated. In fact, six string theories, all with different features, are currently competing within the realm of string theory.

Bosonic string theory, the original string theory, predicts that *bosons* (the force-carrying particles we describe in Chapter 10) are made up of strings. Detractors point to the problem that the theory doesn't predict the existence of *fermions*, which scientists know exist, and also say that *tachyons* (see Chapter 15) do exist, even though no one has ever seen them. Bosonic string theory also needs the universe to have 26 dimensions rather than the familiar 4 of space and time. Luckily, bosonic string theorists say the other 22 dimensions are hidden from view.

The remaining five theories are all so-called *superstring theories* because they require what is known as *supersymmetry*, a theory that the currently known fundamental particles of the Standard Model have heavier partners. These *superpartners* haven't been observed to date because their masses lie beyond the reach of current particle accelerators – but the masses may be within range of the new Large Hadron Collider at CERN, which is due to open in 2008 (turn to Chapter 10 for more about particle accelerators). If superpartners do exist, they solve a number of tricky mathematical problems surrounding the current Standard Model.

The superstring theories solve the fermion and tachyon problems associated with the bosonic string theory and also reduce the number of dimensions in the universe to a mere 10 (rather than 26).

The five superstring theories differ in the following ways:

- Whether the strings are *open* (like a line) or *closed* (like a wobbly circle)
- Whether the theories treat clockwise and anti-clockwise strings differently
- How fermions are allowed to spin
- The maths used to describe the theory

No evidence has yet been found to support any of the superstring theories. Although they offer attractive features – essentially giving a quantum theory of gravity – nothing concrete has been found.

That's not to say that scientists aren't looking. One of the most promising experiments is the AMANDA neutrino detector at the South Pole. This device detects neutrinos that are coming from the skies above the South Pole and also those that have passed through the bulk of the Earth. Scientists have predicted the number of each type of neutrino that should be present. Any discrepancy may provide indirect evidence that one of these string theories is correct.

Finding such evidence is of key importance for cosmologists interested in the early moments of the universe, when the size of the entire universe was extremely small.

Evidence for or against string theory may be the key to unlocking the origins of the universe. Because our current laws of physics don't apply before the Planck instant (10^{-43} seconds after the birth of the universe), string theory offers a way to bridge that gap. Perhaps we can discover whether the fundamental forces were unified, what caused them to separate, and how the early structure of our universe was formed.

Chapter 12

Playing with the Universe's Chemistry Set

In This Chapter

- Exploring the chemical elements that make up the matter in the universe
- Forming hydrogen, helium, and lithium soon after the Big Bang
- Creating heavier elements through nuclear reactions in stars

The universe, as you may have noticed, contains a great variety of very interesting *stuff*. Books, like this one, are an excellent example, not to mention tubs of vanilla ice cream, quasars, giraffes, aircraft carriers, asteroids, party balloons, and so on. Take a look around you now. Stuff is everywhere. The question is, what's all this stuff made of?

At a sub-microscopic level, all this stuff is made up of combinations of different types of atom – otherwise known as chemical elements (which we explain in the section, 'Strolling Through the Periodic Table'). Books are largely made of the element carbon, for example, whereas aircraft carriers contain plenty of the element iron, and balloons are sometimes filled with the element helium, in the form of a gas.

The question of where all these elements come from is a central one in cosmology. In this chapter we explore how the lightest and most abundant elements – hydrogen and helium – formed during the early minutes after the Big Bang (refer to Chapter 6). We also see how the rest of the stuff all around has been generated over billions of years by the thermonuclear reactions that have occurred in many generations of stars.

Strolling Through the Periodic Table

If you did any science classes at school, you have at least a passing familiarity with the periodic table of elements. You may have seen this chart hanging on the wall in the classroom – a grid of boxes each containing a mysterious collection of letters and numbers, similar to Figure 12-1.

1 H																	2 He
3 Li	4 Be											5 B	6 C	7 N	8 O	9 F	10 Ne
11 Na	12 Mg											13 Al	14 Si	15 P	16 S	17 Cl	18 Ar
19 K	20 Ca	21 Sc	22 Ti	23 V	24 Cr	25 Mn	26 Fe	27 Co	28 Ni	29 Cu	30 Zn	31 Ga	32 Ge	33 As	34 Se	35 Br	36 Kr
37 Rb	38 Sr	39 Y	40 Zr	41 Nb	42 Mo	43 Tc	44 Ru	45 Rh	46 Pd	47 Ag	48 Cd	49 In	50 Sn	51 Sb	52 Te	53 I	54 Xe
55 Cs	56 Ba	57 La	72 Hf	73 Ta	74 W	75 Re	76 Os	77 Ir	78 Pt	79 Au	80 Hg	81 Tl	82 Pb	83 Bi	84 Po	85 At	86 Rn
87 Fr	88 Ra	89 Ac	104 Unq	105 Unp	106 Unh	107 Uns	108 Uno	109 Une	110 Unn								
			58 Ce	59 Pr	60 Nd	61 Pm	62 Sm	63 Eu	64 Gd	65 Tb	66 Dy	67 Ho	68 Er	69 Tm	70 Yb	71 Lu	
			90 Th	91 Pa	92 U	93 Np	94 Pu	95 Am	96 Cm	97 Bk	98 Cf	99 Es	100 Fm	101 Md	102 No	103 Lr	

Figure 12-1: The periodic table of chemical elements.

Dry and impenetrable as the periodic table may seem to generations of young school children, it actually encapsulates one of the marvels of cosmology.

- Each of the little boxes in the periodic table represents a *chemical element*, a substance that cannot be further subdivided into more basic constituents by chemical means (such as by pouring acid onto it or reacting it with another chemical).
- Each element is made of one type of *atom*, the most fundamental subdivision of a chemical element which still exhibits chemical properties (such as acidity or alkalinity).
- Each atom is distinguished from the others by one thing – the number of positively-charged particles (or protons) in its nucleus.

We discuss the details of protons in Chapter 9.

Starting with hydrogen

The place to begin looking at the periodic table is at the top left, where the letter H represents hydrogen, the simplest and lightest of elements. As the number 1 in the box indicates, its nucleus contains just a single proton.

You can think of hydrogen as the building-block for all the other elements because, as we discover later in this chapter, the other elements pretty much all result from a process that begins with hydrogen nuclei (in other words, protons) smashing together in the furnaces of stars.

Scan the periodic table and you can see some clues about the importance of protons/hydrogen nuclei. As you read along each row, the number in the box goes up by one. This number, called the *atomic number*, represents the number of protons in the nucleus of the element. So boron (letter B) has five, whereas carbon (C) to its right has six, and one step farther along nitrogen (N) has seven.

Adding neutrons to the mix

All elements other than hydrogen also have another type of particle in their nucleus, called a *neutron*. The total number of these non-charged particles in the nucleus of a particular element can vary. For example, carbon can contain six, seven, or eight neutrons, as well as its six protons.

Adding together the number of protons and neutrons in the nucleus of an atom gives you its *atomic mass*. In the case of carbon, the atomic weights of the different variants are 12, 13, and 14. These variants are called *isotopes*, and can be written in a shorthand as carbon-12, carbon-13, and carbon-14. (Sometimes they're also written as ^{12}C, ^{13}C, and ^{14}C, but we stick with the other form in this book.) Typically, one of these isotopes will be far more common than the others in nature because its nucleus is more *stable* – that is, less likely to decay through emitting particles from its nucleus and turning into something else. The isotope carbon-12, for example, is by far the most abundant in nature, accounting for almost 99 per cent of all carbon.

Watching the clouds

Electrons, the negatively charged particles we met in Chapter 9, really define the science of chemistry. The atom's final constituent is a bunch of electrons that surround the nucleus.

Yet the traditional model of the atom with electrons in orbit – the one that everyone recognises as the symbol of the atomic era – has been shown not to truly represent reality. In fact, electrons act as though they're smeared out in the space around the nucleus in structures that better resemble clouds than well-defined orbits.

Chemistry is thus the science of the interaction between the electron clouds of different atoms.

Calculating the abundance of the elements

Scientists have long known that some elements are more common in the universe than others. Through careful observation of the spectra of light coming from stars, along with other methods, astronomers have shown that by far the most common element in the universe is hydrogen, followed by helium.

In fact, something like 75 per cent of the mass of ordinary matter in the universe is hydrogen, and roughly 25 per cent by mass is helium. (Note, we're putting aside for the moment dark matter, that mysterious stuff we describe in Chapter 11.) All the heavier elements, including the carbon of which humans are mostly made, make up about 1 per cent of the total ordinary matter in the universe.

When scientists examine the 1 per cent of matter that isn't hydrogen or helium, more inequalities emerge. Some elements such as carbon and iron are relatively common, whereas others, like beryllium or gold, are rare. Why is this so? Well, the answer has to do with the way the elements are made. Read on.

Making Helium and Hydrogen in the Big Bang

To chart the history of all the various elements that appear on the periodic table, you need to go right back to the beginning of the universe. By the start, we mean the first few fractions of a second after the Big Bang (refer to Chapter 6), when extreme heat and pressure meant that ordinary matter was a kind of melee of the most elementary particles.

Roughly 10^{-4} seconds after the Big Bang, things cooled down enough (well, to a toasty 10^{12} kelvin at least) for protons and neutrons (which particle scientists collectively call *baryons*) to start forming from the primordial soup. At this stage, an equal number of protons and neutrons existed in the universe.

At this point, the universe was also populated with lighter particles (known as *leptons*), which include such things as electrons and neutrinos. All these particles interact via something called *the weak force*, which allows protons to convert into neutrons and vice versa (we describe the weak force in detail in Chapter 10). With plenty of energy around, as many protons converted into neutrons as neutrons into protons through various particle interactions involving neutrinos, antineutrinos, positrons, and electrons. Yet this situation did not last.

As time passed, the universe rapidly cooled and expanded. Knowing the temperature of the universe at these early stages is vital to our understanding because it governs the amount of energy available for interactions between particles and the creation of particle-antiparticle pairs from photons.

When the universe was a second old, and the temperature had fallen to a mere 10 billion kelvin, the constant flipping between protons and neutrons ended. Why? Well, a neutron has more mass than a proton. When plenty of available energy is around, this isn't a problem. However, at this critical point in the universe's development, the decay of a proton into a neutron became less likely and the protons started to outnumber the neutrons. The decay of neutrons into protons, electrons, and anti-neutrinos became the dominant process and the ratio of protons to neutrons in the universe rose to about six to one.

Between about 100 seconds and 30 minutes after the Big Bang, conditions were just right for protons and neutrons to collide with enough energy to form some of the simplest elements; a process known as *nuclear fusion*. As a result of this confinement within atoms, the neutron decay was halted and the ratio of protons to neutrons in the universe essentially became fixed at seven to one.

After the hydrogen nucleus itself, which is just a single proton, the next simplest atomic structure is called a *deuteron*, which consists of a single neutron and proton bound together and which is essentially the nucleus of the isotope hydrogen-2.

Take a look at Figure 12-2. When other protons (p) or neutrons (n) collided with deuterons (D) with enough energy, they released energy in the form of photons (γ) in the process, and formed helium-3 (which contains two protons and one neutron) and radioactive tritium, the isotope hydrogen-3 (two neutrons and one proton). Figure 12-2 shows some of these reactions.

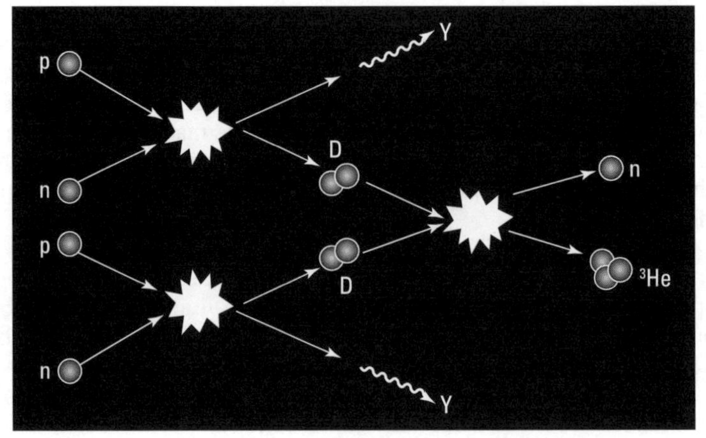

Figure 12-2: Some fusion reactions from the early universe.

Through a variety of different reactions over the next few minutes, collisions between deuterons, helium-3, and other protons and neutrons formed helium-4 (with two neutrons and two protons). Very small amounts of lithium-7 (with four neutrons and three protons) were also created.

Within half an hour, the expansion and cooling of the universe brought these interactions to a standstill. What was left was mostly hydrogen, with quite a lot of helium-4 and smaller amounts of deuterium (hydrogen-2), helium-3, and lithium-7.

Using statistics, scientists have been able to calculate how much of each of these elements the Big Bang should have produced. We mentioned that now 7 protons exist for every neutron or, put another way, 14 protons for every 2 neutrons. A helium nucleus contains two neutrons and two protons. That means 12 spare protons in the universe correspond to every helium nucleus.

Helium should therefore make up roughly 25 per cent by mass of the matter in the universe. In fact, this amount is very close to what scientists find when they look at objects in the universe in which the original abundance values are preserved as nearly as possible, such as dwarf galaxies, which tend to have far fewer stars and be younger than galaxies such as the Milky Way.

The universe was still a very hot place – far too hot for nuclei to be able to capture electrons to form stable atoms. That wouldn't take place until the universe was 380,000 years old (refer to Chapter 6).

By the time the universe was 30 minutes old, its temperature had dropped to 100 million kelvin and there was no longer enough energy to allow nuclei to approach each other closely enough for this nuclear fusion to continue (don't forget, they had to struggle to overcome an electromagnetic force between positive charges!). This means that the proportions of the various elements created in the Big Bang so far remain pretty much constant for the foreseeable future.

Okay, so that explains where all the hydrogen and helium in the universe comes from, but how do the heavier elements – such as carbon and oxygen – arise? The answer is written in the stars. And we don't just mean that metaphorically.

A Star Is Born

Some hundreds of millions of years after the Big Bang, the hydrogen and helium that had formed in the early universe began to aggregate under the pull of gravity, gradually packing together more and more tightly and becoming hotter and hotter – reversing in those areas the general cooling trend of the universe. This point marks the beginning of the first star formation, a process that's still shrouded in some mystery. We explore the formation of the first stars more deeply in Chapter 13.

To get a good understanding of how the various chemical elements on the periodic chart came into existence, you just need to know that as the cold clouds of hydrogen and helium molecules gathered together more closely, the matter at the centre of that gathering gradually became so incredibly hot and dense – more than 10,000,000 kelvin and 100 grams per cubic centimetre – that nuclear fusion was able to take place once more.

The creation of elements inside stars is known as *stellar nucleosynthesis*. In the dense regions that formed as the remnants of the Big Bang coalesced, the first fusion to take place was the combination of two protons (otherwise known as two hydrogen nuclei) to form a deuteron. Under normal circumstances, two positively charged particles repel one another, a little like the positive poles of two magnets. But when protons have enough energy and are moving fast enough (as they do in the high temperature regions at the heart of most stars), they can overcome this repulsion and get close enough for the *strong nuclear force* to bind them together. (We talk more about strong force in Chapter 10.)

The upshot of this collision is the formation of a deuteron, with two protons bound together, plus the release of two other sub-atomic particles, a positron and a neutrino, as well as the release of some energy. This is the same nuclear fusion process that took place in the early moments of the universe.

This process, sometimes called *hydrogen burning*, is actually a very slow one. Although protons are colliding with one another in the core of a star many times every second, scientists calculate that each individual proton only meets and fuses with another on average every billion years. Thankfully, stars have plenty of protons!

Getting from hydrogen to helium

After a deuteron forms, the next step in the genesis of heavier elements happens relatively quickly. Within a second, another proton fuses with the deuteron to produce a nucleus of helium-3.

From here, a couple of things can happen:

- Two helium-3 nuclei collide, forming a nucleus of helium-4 and two free protons.
- One helium-3 nucleus and one helium-4 nucleus (created from the previous reaction) combine to form a heavier element, called beryllium-7.

When an atom of beryllium-7 is formed, two subsequent nuclear fusion processes can occur. The formation of beryllium-7 can do either of the following:

- Capture an electron to become lithium-7, which then collides with another proton, forming two helium-4 nuclei.
- Collide with another proton to form beryllium-8, which can fuse with a positron and become two helium-4 nuclei.

Whether any or all of these nuclear reactions take place depends on the conditions within the heart of the star. If the reactions do happen, they occur at different rates, giving rise to the different proportions of the various elements we observe.

The end result is that hydrogen is turned (or more accurately, the solitary protons that form the nuclei of hydrogen are turned) into the nucleus of helium, which contains two protons and two neutrons.

In these chain reactions, the products of the fusions always have a lower mass than the combined masses of the particles that smashed together in the first place. If we think about this change in mass in light of Einstein's $E = mc^2$ equation – which says that mass and energy are interchangeable (flip to Chapter 4 for more on this equation) – you realise that creating helium out of hydrogen also releases large amounts of energy. In fact, this energy is what helps to stabilise stars, counteracting the effect of gravity and preventing stars from collapsing altogether. This energy is also the source of their light and heat.

Facing a chemical hurdle

In the preceding section, we describe how the earliest stars used hydrogen, the simplest and most abundant element in the universe, as fuel to produce helium and a couple of other elements, such as lithium and beryllium, or more specifically, the isotopes of lithium-7 and beryllium-8.

But a problem exists: Beryllium-8 is actually incredibly unstable (the stable isotope of beryllium found in nature has five neutrons, not four). Beryllium-8 decays back into two helium-4 nuclei in a tiny fraction of a second – seemingly not long enough for it to be any use in building the heavier elements that you see all around.

This major puzzle perplexed scientists for some time. How did the universe make those elements? Particularly, how did the universe make carbon (the life-giving element – at least for life as we know it), which has six protons and six neutrons?

The English astronomer Fred Hoyle revealed the answer in the 1950s. Hoyle was fascinated by the carbon problem. He knew that a way must exist for a third helium nucleus, sometimes called an *alpha particle*, to stick to beryllium-8 pretty much immediately after it formed – a process known as a *triple-alpha collision*.

Why should that be? Well, if the universe had no other way to make carbon, Fred Hoyle wouldn't have been around to ponder its existence, right? That was Hoyle's thinking, anyway.

Getting hotter

The first condition needed for triple-alpha collision to take place is the combination of high temperature and density.

However, stars only get hot enough to make carbon when they've exhausted their supply of hydrogen, because when a star runs low on hydrogen, the pressure in its core drops and the star shrinks under the inwards pull of gravity. This shrinkage in turn creates a higher energy state – the star gets vastly hotter and has greater pressure.

Hoyle realised that another condition is necessary for stars to produce carbon via triple-alpha collision. That condition has to do with the energy of the different elements involved. Hoyle knew that according to quantum physics (see Chapter 9), atomic nuclei normally spend their days in a low energy state (much like some students we know!). This state is called *the ground state*, but occasionally the nuclei take on board some energy (like someone gulping down a can of cola) and enter into an excited state.

Hoyle figured that some situation must exist where the combination of beryllium-8 and helium delivered just enough energy to take carbon-12 to its excited state.

Resonating perfectly

In fact, when he did the calculations – adding the masses of beryllium-8 and helium-4, subtracting the total from the mass of carbon-12, and converting the difference into energy using Einstein's $E = mc^2$, Hoyle came up with a number. He predicted that carbon-12 must have an excited state with an energy level or resonance exactly 7.65 mega-electronvolts (see the Appendix for what this unit of energy means) above its ground state.

When Hoyle first suggested this idea in the 1950s, no experimental evidence suggested that such a situation existed. But Hoyle managed to convince US scientist Willy Fowler and his colleagues at the Kellogg Radiation Lab at Caltech to test his hypothesis using a particle accelerator – and he was soon proven right. Amazing.

Making heavier elements

After a star has produced carbon, the generation of the heavier elements carries on for a while in a relatively stepwise fashion.

To begin with, the star continues burning helium at its core until its supplies of that element are exhausted, at which point the star's core contracts again, making the star hot enough for carbon-12 to combine with another alpha particle to form oxygen-16.

Then the same kind of process that we describe in the earlier section 'Getting hotter' happens again. All the carbon fuel is exhausted, so the core contracts once more, getting hotter, and another round of fusion takes place, producing heavier elements such as magnesium, oxygen, and neon.

In stars that are big enough, this process of stellar burning of elements followed by gravitational collapse is repeated time and again:

- The star burns one element until it runs out in the core (although burning of that element may still occur in outer layers of the star).
- The material making up the star collapses towards the star's core as gravity temporarily takes over.
- The temperature increases.
- The products of the previous fusion processes become the new fuel.

Gradually, in large enough stars, the stepwise conversion of helium to heavier elements continues until elements such as magnesium-24 and silicon-28 are produced. Then, in even hotter and denser stars, the processes become more complex, with elements breaking down and reforming to create ever-heavier nuclei – right up through the periodic table as far as iron-56, which scientists consider the most stable nucleus around with 26 protons and 30 neutrons bound together more tightly than in any other element.

Cycling with carbon, oxygen, and nitrogen

In the Earth's own Sun, which is a moderately sized star, most energy is generated by hydrogen-burning, in which protons crash together as we describe at the start of the earlier section 'A Star Is Born'.

But in larger stars, another process for converting hydrogen into helium dominates. That process is called *the CNO cycle*, named because of the three main elements involved – carbon, nitrogen, and oxygen.

In the multi-step CNO cycle, which Figure 12-3 depicts, carbon-12 collides with a proton to form nitrogen-13. That element then loses a positron to form carbon-13. That element then gains two more protons in stepwise fashion to become oxygen-15, losing another positron to become nitrogen-15, before spitting out a helium-4 nucleus to turn back into carbon-12. Neat, isn't it, the way all the loose ends get tied up, like some finely crafted detective novel? Each step along the way releases energy too.

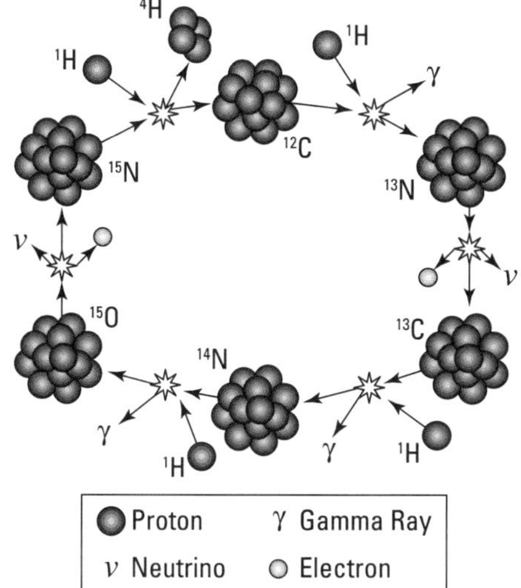

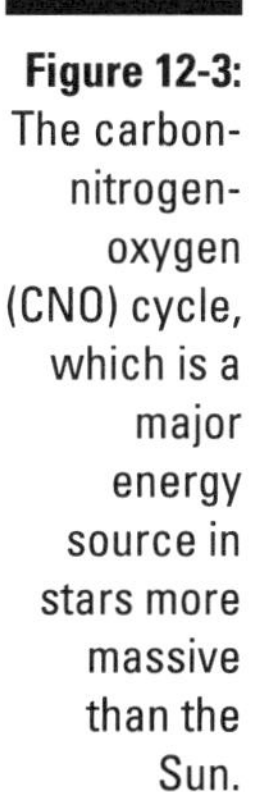

Figure 12-3: The carbon-nitrogen-oxygen (CNO) cycle, which is a major energy source in stars more massive than the Sun.

But hang on a moment. You may find yourself wondering where all that carbon, nitrogen, and oxygen comes from if the star is still burning hydrogen. According to the step-by-step process we describe in the preceding section, 'Making heavier elements', these heavier elements are only created after stars exhaust their supplies of the lighter elements.

Well, the stars you see in our galaxy, the Milky Way, are actually enriched with elements heavier than hydrogen and helium. That means these heavy elements can't have been formed from the baryonic material created in the Big Bang.

Classifying Stars by Their Chemistry

As well as classifying stars by brightness (as we discuss in Chapter 5), astronomers put stars into categories based on their chemical compositions. These categories are called *populations*.

Population I stars

Stars, like the Earth's Sun, are what astronomers call *Population I stars*. They're young stars, mostly found in the discs of galaxies (as opposed to the galactic halos we talk about in Chapter 13), and are rich in elements heavier than hydrogen and helium.

By the way, keep in mind when reading about chemicals in the universe that astronomers refer to all elements heavier than hydrogen and helium as 'metals'. This usage is quite different to the terminology employed by chemists – and, presumably, quite annoying to them as well.

Scientists think that the Sun and other Population I stars didn't form from the material created in the Big Bang. Instead, they formed from material that had already been created through nucleosynthesis by earlier generations of stars and recycled at the end of their lives by means of explosive supernovae (see the later section 'Creating Heavy Metals with Supernovae').

Interestingly, Population I stars are most likely to have Earth-like planetary systems associated with them – because terrestrial planets are formed by the *accretion* (the gravitational clustering) of heavy elements (Chapter 13 describes how planetary systems are formed).

Population II and III stars

When astronomers look into the bulging centre of the Milky Way galaxy, or out to the sparsely populated 'halo' surrounding it, they find another type of star, known as a Population II star. (You can find out more about the shapes of galaxies in Chapter 13.)

The *Population II stars* that are around today are the smaller and longest-lived remnants of the whole Population II clan, dating from as early as 13 billion years ago. The Population II stars are older than the Earth's Sun and contain a smaller proportion of metals. Their bigger, shorter-lived Population II compatriots would have built up heavy elements before going up in supernova smoke and scattering their contents.

But even the Population II stars aren't the original stars in the universe. When scientists look for evidence of metals in their outer layers – where nucleosynthesis hasn't been taking place – they see evidence of heavy elements. This observation means that an earlier population of stars – *Population III* – must have existed and created those elements and scattered them to the winds (so to speak) at their violent deaths.

These early Population III stars are likely to have been enormous – more than tens of times the mass of the Earth's Sun – and short-lived. Scientists think that these stars created and distributed their heavy elements into the *inter-stellar medium* (literally the space between stars) within the first billion years of the life of the universe.

Creating Heavy Metals with Supernovae

The death of stars, a process that in some cases is fantastically explosive, generates enormous energy and scatters much of (or all) the stars' matter through space. Most theories of stellar evolution suggest that it's this supernova material streaming through space that triggers the gravitational collapse of interstellar gas and dust and forms the basis of new stars.

But this kind of explosive ending is a fate reserved for the minority of stars that are significantly more massive than the Earth's Sun. For the most part, as a star burns all its hydrogen into helium and then helium into carbon, it doesn't get hot enough to turn carbon into oxygen. Instead, the star shrinks down on itself, forming what's known as a *white dwarf*, an inert core of matter with the mass of a star crammed into something the size of a planet. When the white dwarf is formed it's still very hot and astronomers observe this fading light when they look through their telescopes. A white dwarf has no source of nuclear fusion to generate any additional light and scientists speculate that it eventually cools down into a cold *black dwarf.*

However, for those stars with masses several times that of the Sun, a more cataclysmic finale awaits – the supernova. Astronomers have now categorised supernovae into two main types, depending on how the stars involved gained the necessary mass in the first place.

Type I supernovae

If a white dwarf has another star in close enough proximity to it, the gravitational pull of the collapsed star can be enough to begin pulling matter off the companion star and wrapping it around the white dwarf itself. If this matter is enough to push the mass of the white dwarf over a kind of tipping point at which its mass is too great for it to withstand the inward pull of gravity, then a supernova is the result.

The critical mass needed for a supernova to happen is roughly 1.4 times the mass of the Earth's Sun – a number known as the *Chandrasekhar limit*, named after the Nobel-winning astronomer who calculated it, Subrahmanyan Chandrasekhar.

After a white dwarf passes this limit, a runaway chain of fusion reactions is triggered, leading to one of the most spectacular events in the universe – a *supernova*. Within seconds, the energy released by the fusion reactions generates an outwardly expanding shockwave, tossing matter outwards at roughly 5,000 to 20,000 kilometres (3,100 to 12,500 miles) per second and creating an enormous blaze of light – up to 5 billion times the brightness of the Earth's Sun.

Supernovae are so bright that the early stargazers thought that they were new stars forming in the sky – hence the name 'nova' for new. The name is ironic, given that supernovae aren't new stars at all, but the dramatic final moments of an old star on its way to oblivion.

Type II supernovae

Another general category of supernova happens to stars of more than roughly nine times the mass of the Earth's Sun. After the various stages of nucleosynthesis have been exhausted, the remaining mass is too great for a white dwarf to form.

In this situation, a core of nickel-iron keeps building up at the star's centre until it reaches the Chandrasekhar limit (see the preceding section, 'Type I supernovae') and collapses, causing protons and electrons to collide and form neutrons and neutrinos.

The neutrons are packed in so tightly that they exert a huge outward pressure, called *degeneracy pressure*.

Degeneracy pressure is caused by the unwillingness of neutrons when they're packed very densely to enter the same energy states as neighbouring neutrons. Quantum physics – and the Pauli Exclusion Principle in particular (refer to Chapter 10) – dictates that particles like the neutron are forbidden from sharing the same quantum state.

As the outer layers of the star fall inwards, they eventually crash into the core and rebound, sending shockwaves out that blow apart the entire star outside the core in an explosion known as a *type II supernova*.

In a galaxy the size of the Milky Way, type II supernovae happen roughly once every 50 years or so. Depending on the original size of the star, they can leave behind black holes (read more about black holes in Chapter 16) or enormous balls of neutrons known as *neutron stars*. These neutron stars are tiny – just over 10 miles across – but contain around 1.4 times the mass of the Sun and are incredibly dense. Gravity on the surface of a neutron star is so strong – perhaps a trillion times greater than that on Earth – that you would be squashed flat in an instant.

Whatever the result, the enormous amounts of energy that a type II supernova releases makes for an environment in which elements heavier than iron can form. In fact, elements up to an atomic mass of 254 are created.

These elements are generated through a complex series of reactions, which we needn't go into in detail. The important point is that the enormous fireball of a type II supernova explosion scatters star dust throughout space. This material (these elements) gives birth to new stars, coalesces and forms planets, and eventually, via the wonders of evolution, combines in the head of a two-legged creature on this particular planet to form the very eyes with which you're reading these words.

Chapter 13

Making Stars, Solar Systems, Galaxies, and More

In This Chapter

- Figuring out the origins of stars, planets, and galaxies
- Comparing today's stars and galaxies to earlier ones
- Numbering the stars – and galaxies and black holes
- Giving the universe shape and structure

As cosmologists try to ascertain how the universe began, one of the biggest questions is how things got from where they were at the time of the Big Bang to how they are now.

The cosmic microwave background radiation that seemingly exists throughout the universe (and which we talk about in Chapter 6) is incredibly uniform, no matter where you look in the sky. This uniformity indicates that in its early days the universe also had very few variations. And yet turn your face to the skies and you see stars, galaxies, dust clouds, and all sorts of very different objects. How can scientists explain the early uniformity and the differences that are evident today?

Additionally, astronomers find themselves in an awkward position when trying to work out how things happen in the universe. Most of the processes involved in building the universe take place over very long time scales – often billions of years. Apart from the occasional stellar explosion, most things in the sky have barely changed in the few millennia that humans have looked upon the heavens in wonder.

Luckily, as we show throughout this chapter, the universe offers a huge numbers of objects for scientists to examine. For example, the universe contains an estimated 62.5 sextillion stars (see the later section 'Estimating the number of stars in the universe' for more on how scientists reached this number). By studying a good number of these celestial objects, scientists can make educated guesses about how stars are born, live their lives, and eventually die.

Building on the basic ingredients – fundamental particles, chemical elements, dark matter, and fundamental forces – astronomers are discovering nothing less than the recipe for making a universe. In this chapter, we explain how stars, solar systems, galaxies, and even larger structures interconnect to form the universe.

Making Stars

When you discover that the Sun is about 4.6 billion years old, you may be tempted to think that all stars have been around for a very long time. This, however, is far from the case. In fact, as you look around the universe, you can find stars being born today. In our galaxy the Milky Way for example, scientists estimate that roughly one star the size of our Sun is created each year (although roughly one star dies each year too).

Stopping by the star nurseries

Star nurseries – large clouds of hydrogen and other gases – yield some of the most stunning images of our universe, as the image of the so-called Pillars of Creation in the Eagle nebula in the colour section shows. These stellar nurseries exist within our Milky Way galaxy. In 2004, for example, NASA's Spitzer Space Telescope found that a distant nebula called RCW49 contained more than 300 newly forming stars.

Although scientists are now pretty sure about where stars are born, they're less certain about exactly *how* they're born. The problem is that stars appear to form from huge clouds of dust and gas, principally hydrogen, measuring tens of light years across. Trying to see anything happening inside these clouds of dust is difficult.

The best theories on the birth of stars suggest that clumps of gas and dust in these clouds congregate together as a result of some disruption, perhaps the close passage of a star or other galaxy, a nearby supernova, or merely the shock of two clouds of gas and dust encountering each other. The result is a spinning disc of gas and dust.

After these initial nudges, gravity is the key force in creating the star. Gravity (take a look at Chapter 3 for more on gravity) acts between the gas and dust particles. This process starts slowly at first, but eventually particles in the cloud tend to move towards more dense areas of the cloud.

This process, known as *gravitational collapse*, increases the cloud's rotation because of the law of conservation of angular momentum. (Think about how

Evolution of the universe. Temperature fluctuations (observed by WMAP) turn into regions of higher density, from which stars form. Filamentary structure then develops until we are left with the universe we see today.

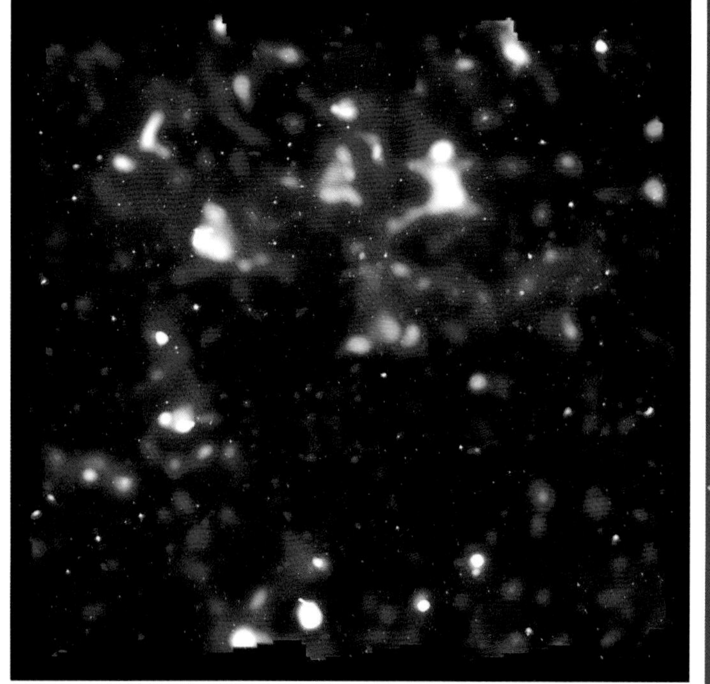

Distribution of normal and dark matter in the universe (red is normal matter, blue is dark matter).

© R. Williams (STScI), the HDF-S Team, and NASA

Looking back 12 billion years: The Hubble Space Telescope's deepest ever view of the universe.

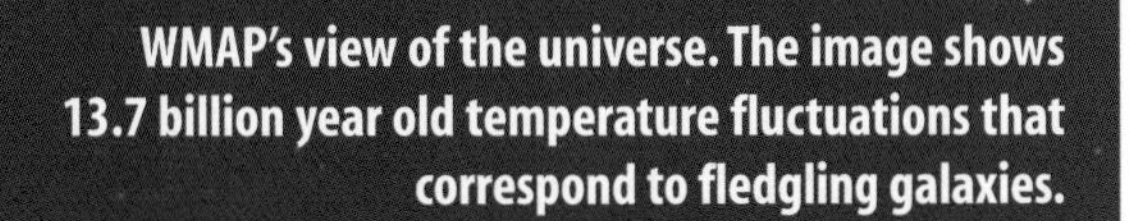

WMAP's view of the universe. The image shows 13.7 billion year old temperature fluctuations that correspond to fledgling galaxies.

© NASA and The WMAP Science Team

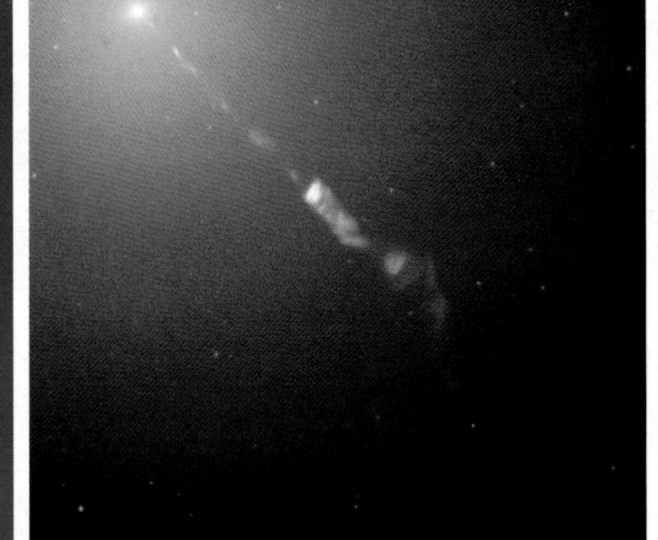

© NASA and The Hubble Heritage Scheme (STScI/AURA)

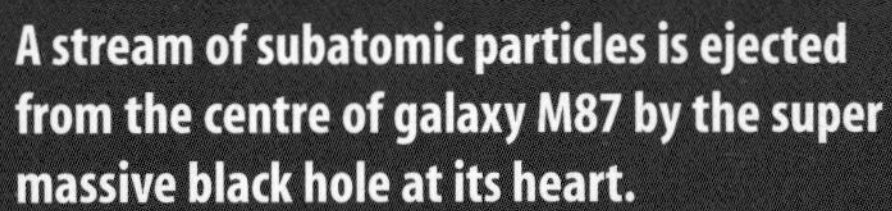

A stream of subatomic particles is ejected from the centre of galaxy M87 by the super massive black hole at its heart.

Spiral galaxies are caught in a cosmic collision by the Hubble Space Telescope.

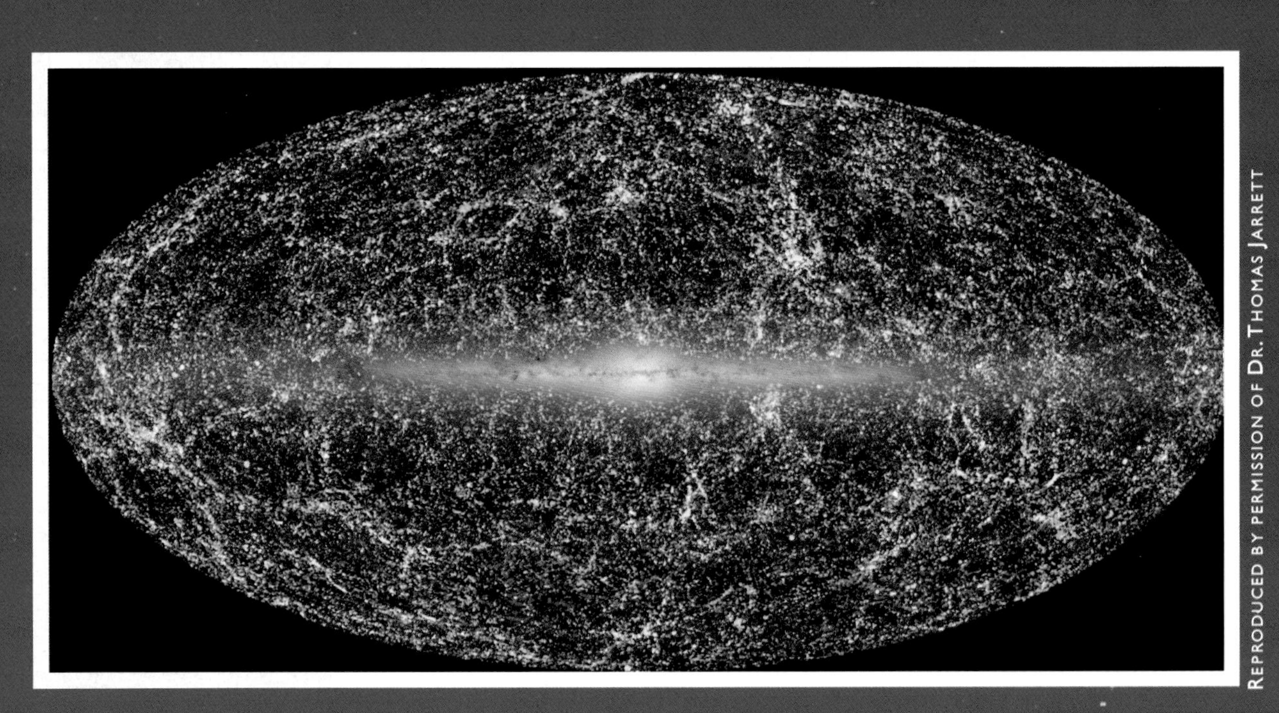

Large scale structure in the universe. At the centre is the Milky Way.

A star nursery in the Milky Way.

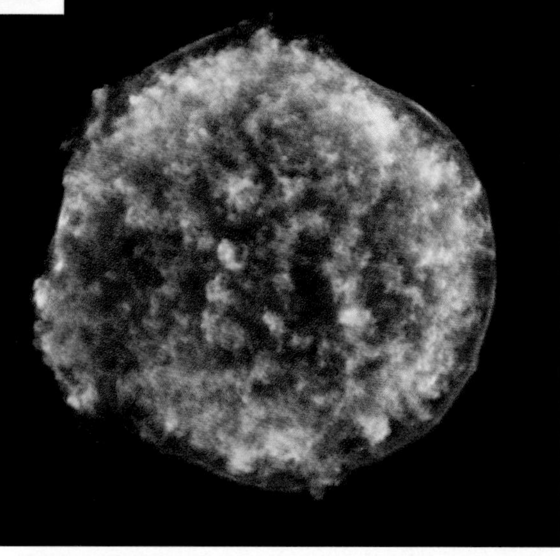

The remnant of the supernova seen by Tycho Brahe in 1572 and imaged by Chandra X-ray Observatory.

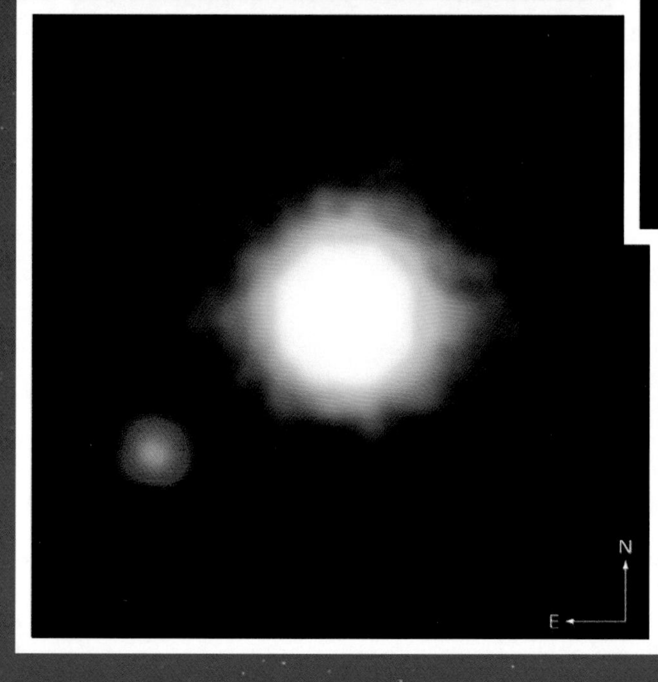

The first image of a planet outside our solar system.

an ice skater spins faster with the arms by the chest rather than with the arms out straight.) In the case of a star cloud, the cloud tends to form into a disc shape, which contains various globules of gas and dust.

The collapse of the cloud also creates heat. The bigger any globule is, the higher its temperature. Eventually, after some tens of millions of years, each globule may have accumulated enough gas and dust to be called a *protostar*. A protostar is not yet a star but is already very hot in the middle. If the temperature within a protostar reaches 10,000,000 kelvin, nuclear fusion can begin, and the star starts shining.

If a protostar doesn't collect enough gas and dust – perhaps the material in the surrounding cloud runs out – it becomes an object known as a *brown dwarf*. These objects are typically larger than the planet Jupiter but smaller than the Sun.

Brown dwarfs are elusive objects, and become fainter and more difficult to see as they age. The first unambiguous detection of one took place in 1995 after decades of searching.

Discovering universal truths from the oldest stars

If you're lucky enough to have a great-grandparent, you're probably fascinated by the stories they tell of life while they were growing up, some time around the beginning of the 20th century. They can speak of life before the Internet, before the jet engine, before television, and before Einstein; a world very different from today. Old stars can tell us a lot as well.

For example, scientists believe that one star in the Milky Way named HE1523 (a boring lot, these astronomers!) is around 13.2 billion years old. As Chapter 6 explains, most scientists now think that the universe is something like 13.7 billion years old. Although HE1523 was only discovered in May 2007, astronomers are already working hard on studying it and similar stars to find out what conditions were like in those very early days.

Despite HE1523's distance, you can discover a lot about the universe's early days from looking at its light spectrum. (Chapter 5 discusses how troughs in stellar spectra are telltale fingerprints of the chemical elements that stars contain.) HE1523's spectrum is interesting because it contains hydrogen and helium, as you may expect, as well as traces of the radioactive elements thorium and uranium. The presence of these metals means that this star was born from the ashes of a *supernova* – the explosion of an earlier star. We can be fairly sure that this earlier star was one of the first ever created in the universe.

In Chapter 12, we explain how chemical elements heavier than lithium are formed in the hearts of stars. When massive stars reach the ends of their lives, the heavier elements are distributed through giant explosions known as *supernovae*. Because these heavy elements can only be created in this manner, scientists have some clues about early stars.

Heavier elements are distributed throughout the observable universe in varying quantities, but all stars seem to contain them. Therefore, the earliest stars must have lived and died very quickly, ending in supernovae that spread the heavier elements far into the early universe, leading to the widespread distribution we see today. To have gone through their life cycles so quickly, these early stars must have been very massive indeed. (The larger the star, the shorter its lifespan because the huge pressures at the hearts of massive stars cause the nuclear reactions to progress more quickly.)

Forming Solar Systems

We talk in the earlier section 'Stopping by the star nurseries' about how a star forms from a disc of material. This disc of material is also what most astronomers believe gives rise to planets.

In addition to the globule in the centre of this spinning disc, which becomes the star, other regions of greater concentrations of gas and dust typically exist within the spinning disc. The effects of gravity cause these areas to attract nearby gas and dust particles, which eventually stick together in clumps known as *planetesimals*. Given enough time, planetesimals collide with others and are thought to be the basic building blocks of what eventually become planets. These so-called protoplanets don't contain enough material to ignite nuclear fusion and become stars. Instead, they're destined to live out their lives as cold lumps of rock or gas, heated by the light from their newly minted star.

The idea of stars and planets being born from a rotating disc of material is known as *the nebular hypothesis* and was suggested as early as the 1730s by Swedish scientist Emanuel Swedenborg.

Some planetesimals never become protoplanets. You can still see the remains of planetesimals littering our solar system today in the form of asteroids, comets, and meteors. The study of *meteorites* – meteors that contain enough material to survive the frazzling journey through the Earth's atmosphere to land on our planet's surface – is a rich source of information on the formation of our solar system. *Comets* – small objects that orbit the Sun – provide even better information about this process, given that they have never had to withstand the searing heat caused by entering the Earth's atmosphere.

Creating Galaxies

Chapter 5 tells how Edwin Hubble was one of the first to classify galaxies according to variations in their shapes – *elliptical* (such as M87) or *spiral* (such as the Milky Way).

Hubble's work also shows that spiral galaxies have different sub-types – some that have central bars through the galactic core and others that just have spiral arms radiating from the core.

Considering spiral galaxies

The spiral galaxies that scientists can observe in the universe reveal several common features:

- A central bulge, or *nucleus*, containing a mixture of young and old stars and, perhaps, a supermassive black hole
- A disc containing younger stars, the spiral arms, and dust clouds
- A spherical *halo*, containing the oldest stars, both individual stars and in globular clusters

Figure 13-1 shows an example of a spiral galaxy with all these features. Also take a look at the image in the colour section of two spiral galaxies colliding.

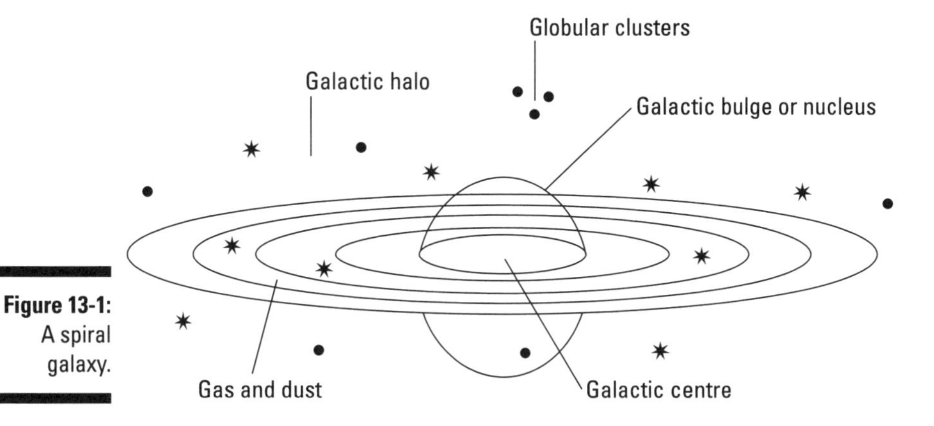

Figure 13-1: A spiral galaxy.

Of course, the preceding observations have to be squared with any theory of galactic evolution that scientists come up with. Not surprisingly, a wide variety of theories exist. However, these theories can be broadly divided into two schools of thought – top-down and bottom-up.

- The **top-down school** believes that the universe was originally a single huge clump of material, which has subsequently broken down into smaller elements.
- The **bottom-up school** suggests that the universe started with fundamental particles (for more on fundamental particles, refer to Chapter 9), which got together to make gas and dust. These particles then clustered together to make the stars, galaxies, and the larger structures.

One widely accepted bottom-up theory involves a heady cocktail of ordinary and dark matter (check out Chapter 11 for more).

The first galaxies are thought to have evolved around 12 to 13 billion years ago – not long after the universe itself formed. Variations in the amount of matter (both ordinary and dark matter) in those early days of the universe dispersed or became more pronounced through the formation of clumps. These clumps, made up of dark matter and ordinary matter, principally hydrogen, eventually formed vast spherical clouds that developed into vast halos.

Because ordinary matter and dark matter interact in different ways (as far as scientists can tell), different things happen to them within the *protogalaxy*, the name given to a cloud that eventually becomes a spiral galaxy. Cosmologists now believe that the ordinary matter falls towards the core of the protogalaxy, forming a halo of old stars made of ordinary matter (perhaps a tenth of the size of a dark matter halo), the nucleus and disc, leaving the halos as home to mostly dark matter. Gravity does the rest.

Forming the spiral arms

The spiral arms of spiral galaxies require additional consideration and explanation.

The big problem for scientists is that if the spiral arms spin round the galaxy and always contain the same stars, they would soon (on an astronomical timescale at least) wind up tightly, the so-called *winding dilemma*. This means we wouldn't expect to see any spiral arm structure these days.

However, this dilemma was resolved in the 1960s when scientists realised that spiral arms are explained using *spiral density waves*. Spiral density waves may seem a complicated subject, but you experience density waves all the time in the form of sound. Sound travels through alternating compressions and rarefactions (the opposite of compression if you're wondering). In air, these variations in density transmit sound from one place to the other – not the movement of air molecules from the source of the sound to your ear.

In a spiral galaxy, the stars, gas, and dust rotate around the core not in fixed circles but in elliptical orbits. These elliptical orbits *precess* (see the beginning of Chapter 4 for a discussion of precession) around the galactic core and lead to variations in density. At the points of higher density, new stars are being born – which explains the appearance of youthful stars in the spiral arms.

Accounting for Everything in the Universe

As well as knowing how the individual components are made, getting some idea of how many of these objects exist can be very useful. If the entire universe consisted of just five galaxies, figuring out how they interacted in the past and formed would be fairly easy.

Unfortunately, the universe contains far more than five galaxies. In the following sections we explain how scientists quantify the number of galaxies, stars, and black holes in the universe.

Clash of the galaxies?

The latest theories argue that elliptical galaxies aren't a stage of evolution in the life of a single galaxy, but arise from collisions between two spiral galaxies. For example, the nearby Andromeda galaxy is approaching the Milky Way. At some point, perhaps 10 billion years from now, the galaxies may collide.

If it happens, the collision is sure to be spectacular, although humanity may not be around to see it. Because galaxies aren't solid objects and contain lots of empty space, the collision isn't going to be just some giant crunch. Computer simulations show that the two galaxies will pass through each other, causing lots of distortion of their spiral structure and some individual collisions. But the galaxies will continue on their way. As soon as they pass through, gravity will start pulling them back towards each other. This yo-yoing will continue for some considerable time (possibly a billion years), eventually resulting in a single huge elliptical galaxy with far less structure than the original two. The cores of the two galaxies will eventually merge through the effects of gravity.

Galaxies

For much of history, most people agreed on the total number of galaxies in the universe – one. Only in the late 19th and the early 20th centuries did astronomers began to question this long-held figure. Specifically, Edwin Hubble's discovery that the fuzzy nebulae were actually galaxies outside the Milky Way inevitably led to speculation about the total number of galaxies in the universe.

The latest best guess on the number of galaxies in the universe comes from the Hubble Space Telescope (HST) – the space-based telescope named after the man who told us so much about the universe.

In 1998, the HST was turned upon a tiny portion of the southern sky in the constellation of Tucana in a ten-day-long experiment called Hubble Deep Field South. By tiny, think how small a grain of sand looks if you hold it at arm's length. All the HST's instruments were focused on this tiny portion of sky, and it picked out an incredible 2,500 galaxies, some extremely faint and others dating back approximately 12 billion years. Scientists multiplied the 2,500 galaxies that the HST observed by the dimensions of the entire sky, and estimate that the universe has a stunning 125 billion galaxies.

The colour section shows an image that the HST obtained by taking multiple exposures and using all the different instruments on board the telescope.

Stars

Just like estimating the number of galaxies in the universe, scientists have to make educated guesses about the number of stars in the universe – counting them is out of the question.

The Milky Way is similar to the billions of other galaxies that populate the universe. This similarity means that if you can work out the number of stars in the Milky Way, you can use this figure to extrapolate to the number in the entire universe.

Instead of taking a representative sample of a small section of Milky Way and multiplying it by the appropriate amount, scientists prefer to work out the total mass in the galaxy. They can come to this figure thanks to Newton's universal law of gravitation.

Think for a moment about the Earth rotating around the Sun. Based on Newton's law, scientists know that a force in addition to gravity must be acting on the Earth. If gravity were the only force acting on the Earth, it would quickly spiral into the Sun and be burned to a cinder.

Grains of sand

Discussions of the number of things in the universe inevitably turn to sand. Why, you ask? The reason is that the numbers involved are so huge that just thinking about them can turn your brain cells to mush.

At some point, some bright spark came up with the idea of comparing quantities to the number of grains of sand on the world's beaches. Although scientists are often sad geeks, no one has yet bothered to count every last grain of sand on the planet. However, they have come up with ways of estimating the number.

The University of Hawaii, for example, worked out that if beaches made up a quarter of all the shores in the world, the average beach was 30 metres wide, the sand was 5 metres deep, and the average grain measured a cubic millimetre, 7.5×10^{18} grains of sand exist on the Earth.

The other force acting on the Earth, a so-called *centripetal force*, keeps the Earth moving along smoothly in a circular path. Because the Earth keeps following the same orbit year in year out, scientists can say that the centripetal forces and gravitational forces are equal. Knowing this, calculating the mass of the Sun is easy. Likewise, by observing the motion of stars around the centre of the Milky Way, scientists can calculate our galaxy's mass.

Using this process, scientists calculate that the mass of the Milky Way is 2×10^{11} times the mass of the Sun. By extension, if all stars in the galaxy are the same mass as the Sun, you can say that the number of stars in the Milky Way is something like 2×10^{11}. But from observations, scientists know that the average star has a mass of about 40 per cent of that of the Sun. As a result, and taking into account the presence of dark matter (see Chapter 11 for more about that), they estimate the number of stars in the Milky Way to be some 500 billion.

Other astronomers measure the luminosity of a galaxy and divide by the average luminosity of a star to get a figure for the number of stars in an average galaxy.

From here, you just need to do some simple sums to get the number of stars in the universe, presuming that every galaxy is like the Milky Way. By multiplying the number of stars in the Milky Way by the number of galaxies in the universe (see the preceding section), you come up with 62.5×10^{21} stars or 62.5 sextillion stars if you're an illionophile – or about 10,000 times more than the number of grains of sand on the Earth (see the sidebar 'Grains of sand' to get a better grasp of this quantity).

Black holes

Counting the number of black holes in the universe has one big problem – the endeavour is like counting the number of black cats in a coalmine.

Although no light can escape from within the event horizon of a black hole (see Chapter 16), a lot of activity occurs just outside a black hole, which gives telltale signs that black holes exist. Collisions within the rotating discs of gas and dust surrounding a black hole give off high-energy radiation that gives them away.

Stars are very often part of a *binary system*, in which two stars orbit each other around their centre of gravity. When one of the stars turns into a black hole, it can start stripping off material from the other star. As matter from one star accelerates into the mouth of the black hole, the stripped away matter gets jostled around by gas and dust particles in the disc and emits high-energy photons, particularly gamma rays. With an appropriate instrument, scientists can detect these photons from the Earth. Scientists have even detected black holes as they are forming.

NASA's Swift satellite, which had instruments on board to detect gamma ray bursts, located more than 200 suspected black holes in the local area (well, within 400 million light years of the Earth, which is pretty local on the scale of the universe).

Many of the black holes that Swift discovered sit in the middle of huge galaxies; others are of the types we discuss in Chapter 16. Some astronomers believe that a black hole sits at the centre of every galaxy, including ours. If this is true, the number of black holes in the universe is at least 125 billion (based on the number of galaxies). However, most scientists believe that black holes only exist at the centre of really huge galaxies. The Milky Way – at 100,000 light years across – falls into this category. As such, perhaps tens of millions of black holes exist in the universe.

Getting the Really Big Picture: Beyond the Milky Way

Galaxies, despite their enormous size, aren't the largest organised structures in the universe. As you can imagine, gravity likes to have a say in the interactions of galaxies.

Peculiar velocities and the finger of God

Although Hubble's Law – which, as Chapter 5 explains, shows that the observed radial velocity of a distant galaxy is proportional to its distance from the Earth, thus causing red-shifting – has been verified again and again by cosmologists, other effects can cause the measured red-shift to differ from expectation.

Galaxies that are part of groups and clusters have velocities within those groups as a result of the mutual effects of gravity. These additional velocities are known as *peculiar velocities* and can sometimes be as much as hundreds of kilometres per second in a seemingly random direction.

For distant galaxies, these peculiar velocities aren't a problem, but peculiar velocity can have an effect on nearby galaxies, known as *the finger of God effect.*

Plotting the red-shifts of galaxies in clusters seems to show that the clusters are aligned in finger-like shapes, and all point back towards the Earth, which some people take to suggest that the Earth holds a special significance and that some greater being wanted to give Earthlings a clue.

In fact, the effect comes down to the peculiar velocities. If they're removed from the calculations, the finger-like nature disappears, leaving a cluster with no specific directionality.

Therefore, galaxies that are near each other – because they were near each other to start with or they passed through another's vicinity – can form small groups. Other groups can then affect these groups, organising themselves into loosely bound structures. In fact, evidence suggests that these structures are everywhere.

Visiting the Local Group

The Milky Way is just one of 40 or so galaxies, including the Andromeda galaxy, which make up something called the *Local Group.* Such groups of galaxies, usually containing a similar number of members, are common throughout the universe and are bound together by their gravitational attraction. This attraction is strong enough to overcome the expansion of the universe and is the reason why, when you look at the light spectra of other galaxies in the Milky Way's Local Group, some galaxies show blue-shifts rather than red-shifts. These blue-shifted galaxies are approaching the Earth rather than moving away from it like distant galaxies that aren't gravitationally bound to the Earth (that is, galaxies where the expansion of the universe overcomes the gravitational attraction).

Clustering galaxies

The next step up in the level of structure in the universe are the galactic clusters, which may contain groups like our own Local Group and consist of hundreds or thousands of individual galaxies.

The nearest big cluster to us on Earth is the Virgo Cluster, which contains more than 2,000 galaxies, and whose centre is in, guess where, the constellation of Virgo (the virgin). It contains several Messier objects (refer to Chapter 5) including the giant elliptical galaxy Virgo A (M87).

Branching out to superclusters

Detailed studies of the distribution of galaxies throughout the universe reveal even larger groupings called *superclusters*, made up of clusters of clusters. The Milky Way's Local Group, for example, is part of a larger galactic get-together called the Local or Virgo Supercluster.

The Local Supercluster is made up of around 100 groups and clusters, including the Virgo Cluster and the Ursa Major group of galaxies, and has the form of a disc with a halo. From watching how these clusters interact, this supercluster is thought to contain as much mass as 10^{15} Suns.

The supercluster that contains the Milky Way isn't the only supercluster. The Perseus-Pisces supercluster, centred on the galaxy NGC1275, stretches some 300 million light years and contains hundreds of groups and clusters.

Scientists believe that superclustering began very early in the history of the universe, although the science of superclusters is still in its infancy. Studies of the distribution of galaxies within superclusters reveal some characteristic features such as *filaments* (galaxies lined up in long strings), sheets of galaxies (detailed in the next section), and galaxy-free voids (like the Boötes void described in the later 'Into the voids' section) between them.

Great Walls

One of the oddest features of the universe is that when you look on a very large scale, galaxies appear to gather in sheet-like layers.

In 1989, two astronomers from the Harvard-Smithsonian Center for Astrophysics announced that they had discovered the largest sheet-like layer in the universe, which has since become known as the Great Wall. Using a database of red-shift measurements, Margaret Geller and John Huchra found a wall of galaxies around 200 million light years away from the Earth. The wall is truly great – measuring some 600 million light years long by 250 million light years wide and only 30 million light years across. Figure 13-2 shows Geller and Huchra's conceptualisation of the Great Wall.

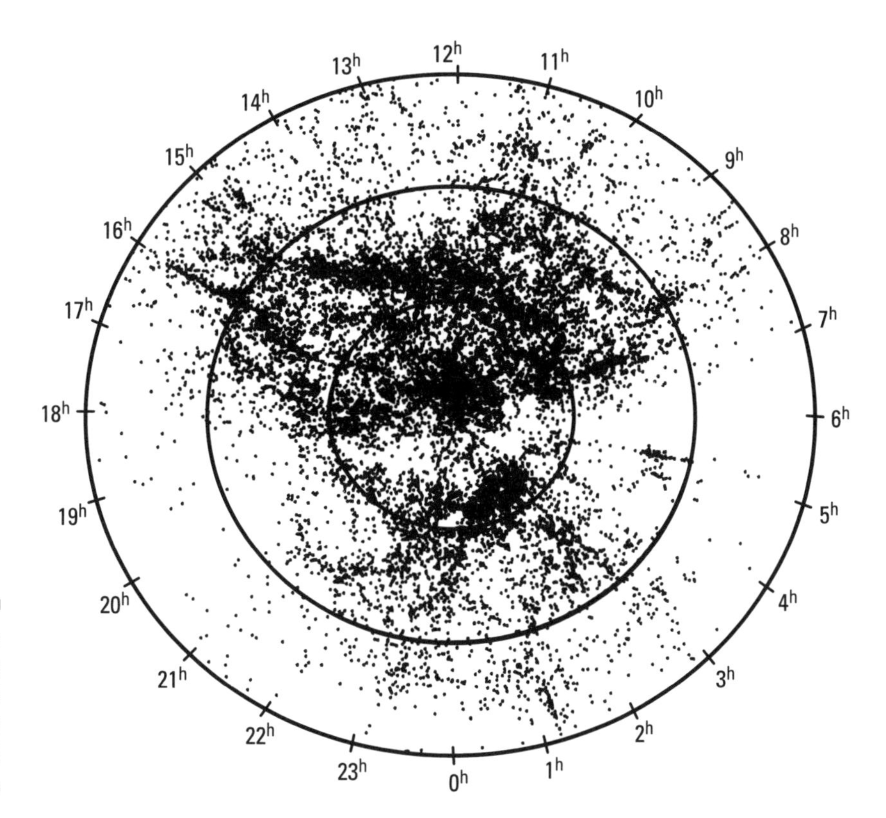

Figure 13-2: The Great Wall © Geller and Huchra)

The intriguing thing for cosmologists is that the huge size of this wall-like structure (the dense area running from 7h to 15h) means that it wasn't formed as a result of the mutual attraction of its constituent galaxies – something external caused the wall to take on its structure. What that something is, however, scientists don't yet know.

The Great Wall isn't just a one-off chance alignment. Although the Great Wall is big, the Sloan Great Wall is even bigger. (The name comes from the Sloan Digital Sky Survey, a project that has mapped a quarter of the entire sky using a dedicated 2.5 metre wide telescope on top of a mountain in New Mexico.) Discovered in 2003, the Sloan Great Wall stretches some 1.4 billion light years and is the most extensive structure yet discovered in the universe.

Into the voids

Although superclusters appear to be linked to each other with string and bridge-like features to make some sort of giant lattice, the same isn't true throughout the entire universe.

In 1981, American-based astronomers Robert Kirshner, Augustus Oemler, Paul Schechter, and Stephen Shechtman announced that they'd discovered a huge hole in the universe. Although not a hole in the ordinary sense, they'd found a region of space measuring 250 million light years across where virtually no galaxies exist. The hole is 700 million light years away from the Earth and located in the constellation of Boötes, the herdsman. If you look at the area with a powerful telescope, you can spot galaxies but they're in front or behind the vast hole.

Explanations of the Boötes void vary. Some scientists say that it occurred through other voids getting together – in the same way that holes in a sock tend to grow by merging with other holes because of the weakened fabric between them. Others argue that the Boötes void isn't a void at all but may contain large amounts of dark matter (refer to Chapter 11 for all about dark matter).

Whatever the explanation, one thing is certain – just like the Great Walls, the Boötes void can't have formed as a result of gravity alone. As the astronomers who found the void say in the scientific paper announcing their discovery: 'It seems . . . that large empty regions have their origins in primordial density fluctuations.' Or, in simple terms, a small region of the very early universe probably contained little matter. The Big Bang and inflation caused this region to grow into a now-huge area of empty space.

Another even larger void, measuring a billion light years across and in the constellation of Eridanus, has just been spotted by astronomers too. It's been dubbed the WMAP Cold Spot.

Weaving a cosmic web

These voids, walls, and other structures (which we don't go into here) come together to create a giant *lattice.* Some theorists liken the universe's lattice-like structure to a mass of soap bubbles, whereas some draw an analogy with biological cells. Others prefer to think of the interconnectedness as some gigantic web.

Humans are in the very early days of understanding about the large-scale structural elements in the universe, and scientific opinion is divided. The top-downers (see the earlier section 'Considering spiral galaxies' for more) believe that this cosmic web is the remnant of what existed in the universe from the very beginning. The bottom-uppers, on the other hand, think that the cosmic web is what the galaxies and clusters are moving towards.

Whichever view is eventually proven right, the fact is that this interconnected structure to the universe exists. Studying this structure is sure to reveal much about the nature of the universe in the years to come.

Chapter 14

Giving Birth to Life

In This Chapter

- Describing life on the Earth – and possibly beyond
- Considering where, when, and how life began
- Pondering the fact that the universe is so well suited to life

Is life important – on a cosmic scale? Because scientists no longer believe that the little blue planet you live on is the centre of the universe, you may wonder why we even include a chapter on life – on the Earth and possibly beyond – in a book about cosmology.

Many scientists say that life is a fluke occurrence, unimportant in the cosmic scheme of things. The physicist Stephen Hawking (author of the infamously unread bestseller *A Brief History of Time*) sums up the argument against life's importance as follows: 'The human race is just a chemical scum on a moderate-sized planet.' Crikey. Perhaps humans should pack their bags and slink back into the slime right now!

But hang on, other scientists have a different take on the matter of life. They point out that if the universe had been set up even slightly differently, if any of the tiny calibrations of the cosmos were just a fraction off, life as you know it would never have come into existence. Some even suggest that the emergence of life is something built into the fabric of the universe.

Read on to hear the arguments for and against the importance of life in the universe.

Defining Life

At first glance, distinguishing between living and non-living things seems like a piece of cake. You probably do it all the time. Strolling down a street, for example, you know instinctively that the phone box on the corner, the streetlights, and the sheets of newspaper blowing in the wind aren't alive. On the other hand, your aunt's vicious tabby cat is very much alive, as are the sparrows it's stalking, and the grass the birds are pecking around in.

But scientists have found that grasping a scientific definition that can draw a line between the living and the non-living is like trying to get hold of a bar of soap in the bath. They know it exists, but the darned thing just keeps slipping out of reach. Think about what makes humans alive – breathing oxygen for example. Perhaps that should be the definition of life? Yet biologists have discovered certain bacteria that are *anaerobic* (don't need oxygen to thrive), yet they're certainly rather lively. Perhaps life is being capable of thought? Nope. Plants don't think – at least not as we know it – yet they're alive too.

Nevertheless, living things do have some characteristics in common – things that living things 'do', so to speak. The following sections describe these characteristics as a way of getting closer to a definition of life.

Complexity

Living things are complex and highly organised. Even the simplest, single-celled bacteria consist of millions of components, all working in synchrony to keep the little creatures functioning.

However, many non-living things are also complex. For example, a galaxy isn't living, but has millions of different components moving around each other in complicated ways.

The real hallmark of the complexity of a living thing is its *organisation.* In his book *The Origin of Life*, physicist and astrobiologist Paul Davies points out that the different elements of a living thing need to work together, or the organism ceases to function.

Metabolism

All living things process chemicals through a set of complicated steps that allow them to extract the energy they need to do things like move around. This process of making energy from chemicals is called *metabolism*, and it's a process you undertake every time you bite into an apple or tuck into a nice plate of egg and chips.

The famous early 20th-century physicist Erwin Schrödinger touched on something related to these chemical processes when he tried to define the fundamentals of living in his book *What is Life?* He said that life was defined by the way living things create order from disorder. Metabolism is the way living things extract order from the environment.

Development

Another aspect of living is the tendency of organisms to change: Animals grow, ecosystems spread, and the grass on your lawn needs cutting every now and then.

Of course, non-living things, such as crystals, also grow and develop over time given the right conditions, so development alone isn't enough to define life.

Autonomy

Physicist Paul Davies points out that having a life of one's own is pretty important to living things.

For example, humans can be subject to the laws of physics – a strong wind can blow you around, gravity keeps your feet on the ground, and you float in water. But you also have the capacity to act of your own volition. If you swim to the bottom of a pool in search of dropped coins, you're proving beyond all doubt that you're alive – as long as you return to the surface before you need to take a breath.

Reproduction

One of the most important defining characteristics of life is reproduction. Birds do it, bees do it . . . even sentimental unicellular organisms do it – after their own fashion. Not only do all living organisms make copies of themselves, but the copies must also include the instruction set for creating copies of their own.

Even this seemingly innocuous statement has its problems, though. Think of a mule, for example. Mules are the result of a horse and a donkey mating. They're definitely alive, as anyone who's been kicked by one can tell you. But mules are nearly always sterile and thus can't reproduce.

Still, reproduction is clearly important in a general sense for living things. You certainly wouldn't be here without it. And as scientists have thought more about life, they've also come to consider as important the natural variations that occur as living things reproduce.

What we're talking about here is *evolution*, first described by English naturalist Charles Darwin (1809–1882). When living things reproduce, their offspring exhibit variations between one another. If those variations make one individual better equipped to survive in the world, that one is more likely to survive and pass on its characteristics to future generations.

Life, but not as we know it

Fans of the original *Star Trek* television series may be familiar with a famous phrase, uttered by Doctor McCoy, the surgeon of the *Starship Enterprise*, when confronted with some alien life form on a distant planet: 'It's life, Jim, but not as we know it.'

The idea that some life forms may exist in space – or on the Earth – that don't match the current definition of life is intriguing. This life form isn't simply something that looks a little different, but whose basic characteristics are different. For example:

- All known life forms on the Earth need water to survive. But need this be the case for all possible forms of life? The answer, laboratory experiments suggest, is no.
- All known life forms on the Earth are built of chemicals that have the element carbon in them. But can other chemicals work as effectively? Maybe.
- Life forms on the Earth have as the instruction manual for building life the nucleic acids RNA and DNA. Can a different system than using these molecules possibly exist? Well, perhaps.

Answering any of the preceding questions requires a fair amount of speculation, but here's one thought to take away. The most important elements in life on the Earth are

- Hydrogen
- Oxygen
- Carbon
- Nitrogen

By contrast, the most abundant elements in the universe, in order of their abundance, as far as scientists know are

- Hydrogen
- Helium
- Oxygen
- Carbon
- Neon
- Iron
- Nitrogen

Considering helium and neon are largely inert, the similarity between the elements of life on the Earth and the elements found most commonly in the universe is pretty remarkable. For some scientists, this similarity suggests that any life forms elsewhere may be made of similar stuff to Earthbound life forms.

Drafting a working definition

Taking all the elements we describe in the previous sections, and all their shortcomings, into account, we can come up with a simple, one-sentence definition of life. (Of course, scientists tell you that this description isn't perfect, but it serves as a rough definition, at least for now.)

Here's the working definition, which is sometimes known as the *NASA definition of life*: Life is a self-sustaining chemical system capable of Darwinian evolution.

Tracking the Very Beginnings of Life

With a working definition of life (see the preceding section), we can push on into even murkier waters – the origin of life on the Earth.

So when did life begin on the Earth? The short answer to the question is that nobody knows. But you didn't open this chapter and start reading this section for a two-word answer. Fortunately, we offer a host of interesting suggestions about how life may have started.

Calculating the age of life on the Earth

To begin figuring out when Earthbound life began, science offers an obvious upper limit. Seeing as the planet is roughly 4.5 billion years old, you can safely say that no life existed here before then. (For more on the formation of planets, see Chapter 13.)

The best place to look for evidence of life after that point is in the rock-bound fossil record. The oldest rocks on the planet date to about 3.9 billion years ago. Earlier in the Earth's history, cosmic debris was constantly bombarding the planet. The Earth's surface was molten and these very old rocks would have been messed around too much by subsequent heat and pressure to preserve fossils.)

Life, however, seems to have appeared pretty quickly after the Earth's molten surface cooled a bit. The Pilbara region of western Australia offers some of the oldest fossil evidence for life on the Earth. Scientists have identified formations called *stromatolites*, which mats of algae may have formed. This evidence points back to 3.5 billion years ago, although in recent years significant debate has ranged about whether the stromatolites were really formed by algae.

Less controversial evidence comes from slightly younger Canadian fossils, which carbon-dating shows to be around 3 billion years old.

Scientists have other methods at their disposal to spot signs of ancient life in rock. One of these carbon dating techniques measures the ratio of carbon isotopes (see Chapter 12 for more on isotopes) in bands of rock. This technique suggests that life may have been around as long as 3.8 billion years ago.

Probing the origins of life

If you work from the basis that life seems to have been around on the Earth for something like 3.5 billion years, living things emerged relatively soon after the Earth was formed.

At that stage, the planet was very different from the one you call home today. Back then, the planet was regularly convulsed by volcanic eruptions, bombarded with meteors, wracked by titanic storms, and strafed by ultraviolet light and lightning. Sounds lovely, huh?

Yet amid this diabolical setting, life somehow emerged from an environment rich in *organic molecules* (compounds containing the chemical element carbon), but with very little oxygen. The question is how?

Pondering panspermia

Some scientists think that life may not in fact have begun on the Earth, but that it was delivered here, perhaps by an asteroid or comet carrying primitive cells or bacteria.

This concept, called *exogenesis*, relates to a wider idea called *panspermia* – which means 'germs everywhere' – and suggests that the wider universe is teeming with life.

Supporters of panspermia, however, face the tricky question of the unlikelihood of life making an extended journey through space, exposed to radiation and cosmic rays, and then surviving the extreme heat and force involved in crashing into the Earth.

On the other hand, meteorites discovered on the Earth show intriguing hints of life. In 1984, for example, US scientists discovered a meteorite from Mars, ALH84001, which they subsequently found contained tiny grains that look very much like tiny bacteria. Scientists still haven't reached consensus about whether Martian life forms made these grains and the organic matter on the meteorite.

As the astrobiologist and physicist Paul Davies writes in his book *The Origin of Life*, Mars, the Earth, and other planets regularly exchange meteorites, which suggests life may have moved between them. Davies notes:

> *It wouldn't take much to convince me that ALH84001 contains genuine fossils because I think there almost certainly was life on Mars 3.6 billion years ago . . . The reason I am so confident in this belief is not because I am sure life emerged from a primordial Martian soup (though it may have), but because the planets are not, and never have been, quarantined from each other.*

Finding organic matter in space

Even if life itself didn't travel to the Earth from outer space, perhaps the building blocks for life did. Since the 1930s, astronomers have been using infrared and radio wave detection to identify molecules in clouds of dust and gas in space. The first molecules found this way were simple combinations of carbon and hydrogen, and carbon and nitrogen.

Since then, ammonia (NH_3), water (H_2O), and formaldehyde (H_2CO) have been spotted, as well as more complex molecules. In 2005, for example, NASA scientists using a space telescope to examine a galaxy 12 million light years away found copious evidence for the presence of 'polycyclic aromatic hydrocarbons', a type of flat, chicken-wire shaped organic molecule formed in the winds of dying stars and spread through space (see Figure 14-1). These molecules are found on Earth today and are considered a possible constituent of the *primordial soup*, the rich watery goo that some scientists believe may have been where life first began, nudged into doing so by intense solar radiation and the odd spark of lightning.

The study's leader, Doug Hudgins, said that the study had implications for where life on the Earth came from in the first place:

> *There once was a time that the assumption was that the origin of life, everything from building simple compounds up to complex life, had to happen here on Earth . . . This stuff contains the building blocks of life, and now we can say they're abundant in space . . . And wherever there's a planet out there, we know that these things are going to be raining down on it. It did here and it does elsewhere.*

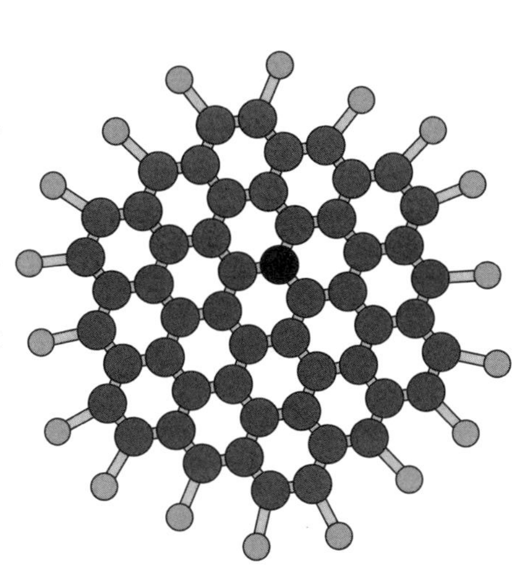

Figure 14-1: A polycyclic aromatic hydrocarbon. The dark grey balls are the carbon atoms, the light grey balls represent hydrogen atoms, and the black ball shows the position of a nitrogen atom within the molecule.

Enjoying a Warm Bowl of Primordial Soup

Organic molecules, such as those found in interstellar dust clouds or common everyday compounds such as ethanol and plastics, are fascinating, but they're not alive. Putting aside for a moment the idea of panspermia – see the earlier section 'Pondering panspermia' – you're left trying to imagine how the fantastic variety of life on the Earth, including humans of course, emerged from a mixture of non-living chemicals.

Scientists think that every life form you're familiar with – from giraffes to amoeba – share a common ancestor. If you trace your family trees back far enough, everyone has the same origin. This notion exists for good reasons.

- All living things use the same molecular systems for carrying genetic information (see the later section 'Getting inside the cell: DNA and proteins').
- All organic molecules have a handedness.

By *handedness*, scientists mean that in nature, many molecules can be constructed in two different ways, which are mirror images of each other. Their chemical structures are the same, but they're reversed right for left and left for right.

But in biology, the option of handedness – or *chirality* – doesn't apply. Molecules are consistent across all life. So DNA, for example, almost always coils to the right, even though nothing in the laws of physics says it must. This consistency suggests that all life evolved from a single ancestor whose molecules had the particular chirality that everything – including you – inherited.

Dipping your toe in Darwin's pond

Charles Darwin, the father of evolutionary theory, once suggested that life may have emerged from non-living matter in 'some warm little pond, with all sorts of ammonia and phosphoric salts, light, heat, electricity'. Given enough time, Darwin thought, life may develop through a process of increasing complexity.

Over the years, scientists have proposed a lot of alternatives to Darwin's shallow pond and its ingredients.

- In the 1920s, John Haldane suggested that life began in an ocean rich in chemicals that had 'reached the consistency of hot dilute soup'.
- More recently, scientists have proposed that the dramatic environments of deep sea hydrothermal vents may have been a possible site for life to begin.

Test tube life

Some of the most famous experiments in evolutionary biology (the study of the origin and change of species), have involved trying to recreate the conditions under which life may have arisen more than 3 billion years ago.

For example, in 1953, Stanley Miller, a student at the University of Chicago, with guidance from Professor Harold Urey, filled a glass flask with a mix of water, methane, hydrogen, and ammonia and passed an electric spark through it to simulate lightning. Over the following week, they watched the water turn a deep reddish colour. It turned out that he'd created several chemicals called *amino acids*, which are the building blocks for proteins. At first, the results were hailed far and wide as the first step toward the creation of life, but scientists soon realised that the Earth's early atmosphere probably didn't look like the gassy mixture he'd used. When Miller later tried again with a more appropriate mixture, the broth failed to materialise.

In early 2007, Jeffrey Bada from the Scripps Institution of Oceanography in La Jolla, California, revisited Miller and Urey's experiment. Bada had discovered that the reactions in the failed experiment produced chemicals called *nitrites*, which destroy amino acids. The nitrites also turned the water acidic, which prevents amino acids forming. Bada added chemicals to the experiment that duplicated the role of iron and carbonate that would have neutralised nitrites and acids in the primitive Earth. The result? The same watery liquid as Miller obtained in 1953, but this time Bada's fluid was packed full of amino acids.

The idea that life emerged from a primordial soup lives on, although more experiments are required to provide a definitive answer.

Getting inside the cell: DNA and proteins

Living organisms are made up of *cells*, small compartments about a hundredth of a millimetre in size. Cells, as Figure 14-2 shows, contain an intricate chemical machinery performing multitudes of tasks at any moment.

The dominant ingredient in the buzzing microcosm of the cell is a class of molecules called proteins. *Proteins* are the workhorses of cells, carrying out most of the major functions that underpin life. Many proteins facilitate the chemical reactions that underlie the processes of metabolism, through which cells get energy from their environment.

Proteins are constructed from long chains of molecules called amino acids. In fact, the millions of proteins found in the Earth's endless variety of living things are all made up of combinations of the same 20 amino acids.

In cells, proteins are made by an intricate molecular machine called a *ribosome*, which constructs the proteins from free amino acids, according to instructions set out in the cell's genetic code. Figure 14-3 shows one of the amino acids that this process strings together.

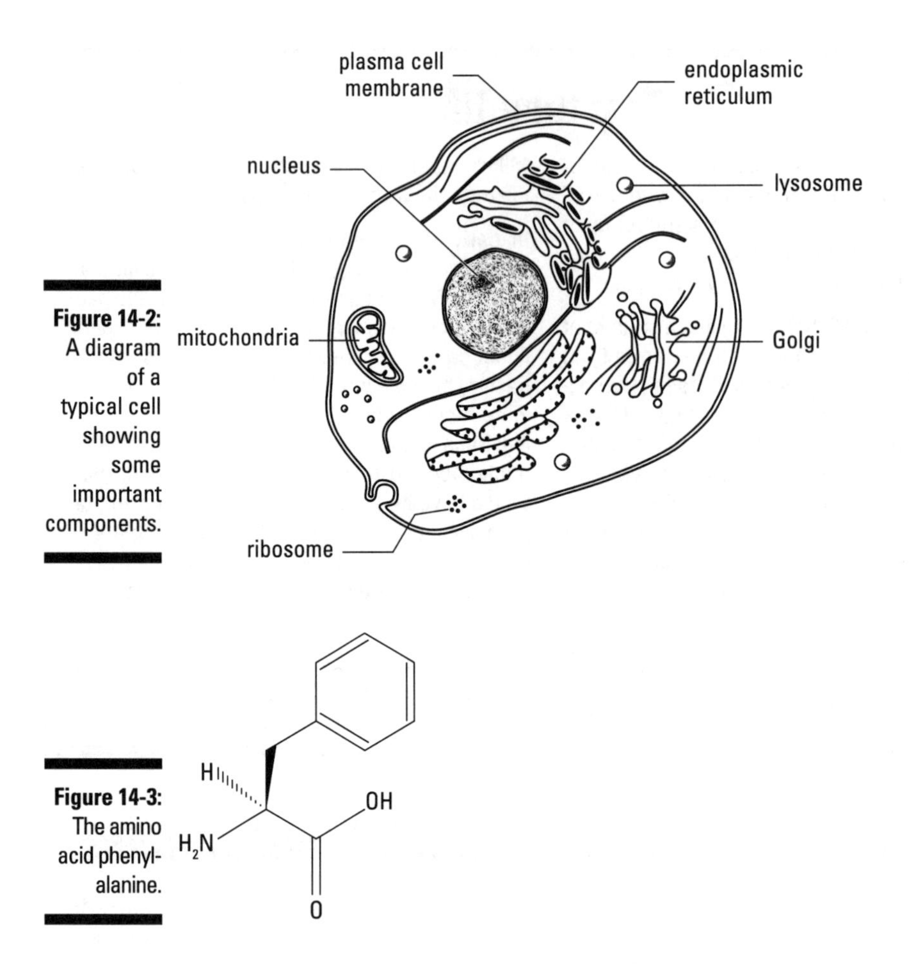

Figure 14-2: A diagram of a typical cell showing some important components.

Figure 14-3: The amino acid phenylalanine.

Amino acids are one of the basic building blocks of life and another is a group of molecules known as *nucleotide bases*. These five different chemical entities are combined to make the nucleic acids DNA and RNA. DNA and RNA are the molecules that carry the instruction manual for manufacturing everything in the cell, including proteins.

DNA is a long molecule shaped like a *double helix*. To visualise the molecule, imagine taking a toy ladder in your hands and twisting both ends in opposite directions so that the long uprights of the ladder become twisted around on each other. The important parts of the DNA molecule are the rungs of the ladder. Each of these rungs is constructed of pairs of nucleotide bases, using four of the five known bases – adenine, cytosine, guanine, and thymine. (The fifth base, uracil, is used to make RNA, another nucleic acid.) Figure 14-4 is a diagram showing the structure of DNA.

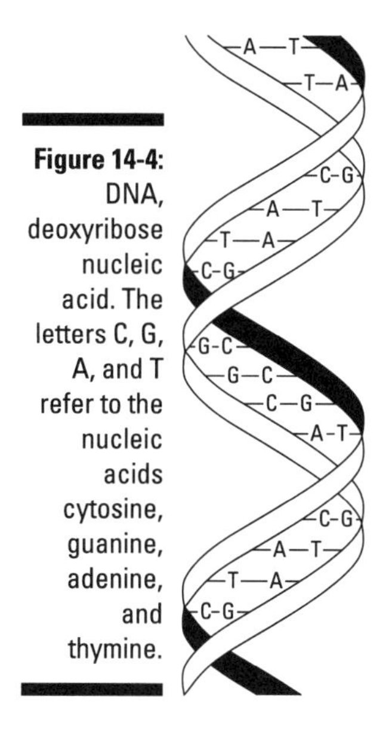

Figure 14-4: DNA, deoxyribose nucleic acid. The letters C, G, A, and T refer to the nucleic acids cytosine, guanine, adenine, and thymine.

DNA is an ingeniously clever molecule. Its design not only carries the information for life, but also allows for easy replication of itself. Just before a cell divides, its DNA molecule divides up the middle of the ladder, splitting the pairs of nucleotide bases apart.

With the help of protein catalysts, each of the half-DNA molecules becomes a template to build a complete replica of the original, so that where one DNA molecule had existed, now two exist – one for each cell.

Because cytosine can bond only to guanine, and adenine only to thymine, the two ‘offspring’ replicas are identical to the original ‘parent’ DNA molecule. This system is vital for reproduction, because the molecule contains all the crucial information about the functions of that specific type of cell.

Anyway, we’ve taken you a long way from cosmology with all this description of DNA and proteins – but we’ve done so for a reason. The interconnectedness of these DNA and proteins cuts to the heart of explanations for the origins of life. Read on.

Addressing the chicken and egg problem

When you think about the proteins and nucleic acids we describe in the preceding section, you can clearly see that neither is any use without the other in terms of sustaining life. They need each other.

One of the major areas of discussion currently among scientists interested in early life is over which of these two pillars of life came first: Did life begin with networks of metabolic reactions – or with a complicated replicator molecule such as RNA?

- **Scientists who favour the metabolism-first idea,** broadly speaking, think that the development of life began with the formation of compartments that contained mixtures of non-living chemicals. These compounds went through a series of reactions, which became more and more complicated and eventually made the leap (somehow!) to storing information in polymers such as nucleic acids.
- **Scientists who support the 'RNA world' hypothesis** suggest that in the early Earth, life formed from a soup of RNA molecules that had assembled by chance out of nucleotide bases. These molecules served as catalysts for their own replication. As molecules made copies of themselves, mutant versions underwent a kind of evolution, eventually leading to the development of cell-like compartments and metabolism.

As things stand, some of these questions about the origin of life are simply unanswered. Whichever scenario scientists invoke, chance is mentioned as an important factor – the chance formation of long nucleic acid chains, the chance formation of proteins, and so on.

The astronomer Fred Hoyle illustrated his scepticism of *abiogenesis* (the word scientists use to talk about life arising from non-living things) in his book *The Intelligent Universe.* He likened the spontaneous assembly of life to a whirlwind passing through a junkyard and forming a fully assembled 747.

Yet you are clearly here. How lucky is that? For some researchers this dilemma raises an interesting point – the universe seems to be perfectly designed for life, as we address in the next section.

Living in a Universe That's 'Just Right'

When Goldilocks stumbles into the cottage of the three bears, she finds that the porridge belonging to the baby bear is 'not too hot and not too cold, but just right' for her needs.

Have humans, like that fairy tale heroine, landed on their collective feet in a universe that's 'just right' for life? More controversially, does an underlying principle in the way the universe is built somehow require the emergence of life?

Well, certain parameters in the make-up of the universe would rule out life if they were slightly different. For example:

- If the strong nuclear force were just a little stronger, the Sun would burn all its fuel in less than a second; not enough time for life on the Earth to develop.
- If the energy state of carbon-12 – which allows the nuclear reactions within stars to generate carbon and heavier elements (refer to Chapter 12) – were slightly different, humans wouldn't be here.
- If dark energy, the mysterious cosmic antigravity force that's pushing space itself apart at an accelerating rate, were only slightly larger in magnitude, galaxies would never have formed.

Is this a cosmic fix? Many scientists say no. For them, remarking on the fact that the universe is fit for life is a little like a puddle waking up one morning and being amazed at how perfectly the pothole fits it. Humans are here on the Earth because it's possible for them to be here, they may say.

Some people in the cosmological world, however, disagree. Their views fall under the umbrella of what's called the *anthropic principle*, a concept formalised in the early 1970s by British researcher Brandon Carter. Carter originally raised the anthropic principle at a meeting to celebrate the 500th birthday of Copernicus, the man who first postulated that the Earth (and by extension humans) don't hold a special place in the universe (flip to Chapter 2 for more on Copernicus).

Carter's initial argument was that the Earth and human beings *do* hold a special place and time. That is, the conditions happen to be just right for the existence of life on the Earth at the present time because if they weren't, humans wouldn't be here.

Since then, the moniker of the anthropic principle has expanded further, to include discussions of whether humans just happen to live in one life-friendly universe among many, or whether only universes that have the potential for life and conscious thought really exist.

If you're finding these ideas mind-boggling, and perhaps questionable, you're not alone. But the whole area is certainly fun to think about.

Chapter 15

Travelling Through Time

In This Chapter

- Probing the scientific meaning of time, space, and spacetime
- Considering the possibility of time travel
- Looking back to a time prior to the Big Bang

In our discussions of the Big Bang and the origins of the universe in the rest of this book, we talk in passing about time. In this chapter, we try to understand exactly what we mean by that word 'time'.

The problem is that since Einstein started fiddling around with the ideas of the universe (as we discuss in Chapter 4), science's concept of time has had to become more sophisticated. Einstein realised early on in his work that time is not fixed at all but instead depends on how fast you're moving relative to the speed of light. This chapter explores how scientists work with this strange concept.

We also talk about the speed of light and consider whether time travel is possible, by looking at weird particles called *tachyons* that some people believe travel faster than the speed of light.

Finally, we examine one of the most difficult-to-comprehend ideas about the origin of the universe: How did the Big Bang happen in the first place? If the universe as humans know it didn't exist *before* the Big Bang, how can anything have suddenly happened? (Some scientists have tried to get around this idea by suggesting that the universe has been in existence forever. Nice try, but we offer some very good reasons for why this isn't the case.)

Exploring Past, Present, and Future

In the everyday world, the concepts of past, present, and future are fairly obvious: You ate that huge cream cake in the past, you're now feeling the sensation of indigestion, and a week in the future you'll feel regret as the bathroom scales reveal that you've ruined your diet.

But think about this. The sunlight that's reaching the Earth at this moment has taken eight minutes or so to travel on its 150 million kilometre (93 million mile) journey. If the Sun decided, for some reason, to turn green and shine green light, it would take you eight minutes to realise what had happened. However, any creatures sitting on the surface of Venus would have noticed the colour change after only six minutes, reflecting the shorter journey time of light between the Sun and Venus.

The light from the supernova seen in a dwarf galaxy called the Large Magellanic Cloud in 1987 took 170,000 years to reach the telescopes of the Earth's astronomers, because the star that emitted the light is 170,000 light years away. Any residents of a planet in orbit around the star that exploded have long since perished. This example shows the complicated nature of answering the question: Whose view of 'now' is right?

Thinking differently about spacetime: Worldlines and light cones

In Chapter 4, we discuss the concept of *spacetime*, that old favourite of science fiction writers. By defining events in terms of their three spatial coordinates (left to right, up and down, and front to back) and a time coordinate, you start to pinpoint events rather than mere objects at certain locations.

This combination of spatial coordinates and time is particularly handy when you're trying to get your head round the notion that sunlight reaching our eyes on Earth was emitted eight minutes earlier. In more scientific terms, you may say that the Sun emits a photon of radiation (see Chapter 9) from a certain point at its surface at a given time and you detect it with your eyes (and perhaps a telescope) at a certain location at another time.

Einstein formulated the idea of *worldlines* as a way of using these coordinates to map the history of an object through both space and time.

Rolling a ball

Consider the following simple example: Imagine that you roll a ball along a straight line. You have a means of measuring its position along the line every second. Figure 15-1 is a graphic you draw to show this motion.

In Figure 15-1, the black dots represent your measurements of the ball's position. Because the ball is moving with a constant speed, the dots are evenly spaced, and you can draw a straight line through them. Think of the thick black line connecting the various dots as the *worldline* of the ball.

The Earth's orbit around the Sun

Consider the Earth going around the Sun. How would you draw a diagram of that? If you assume that the Sun is fixed in position and that the Earth's orbit is flat, you can draw the diagram using only two space dimensions and one time dimension, as Figure 15-2 shows.

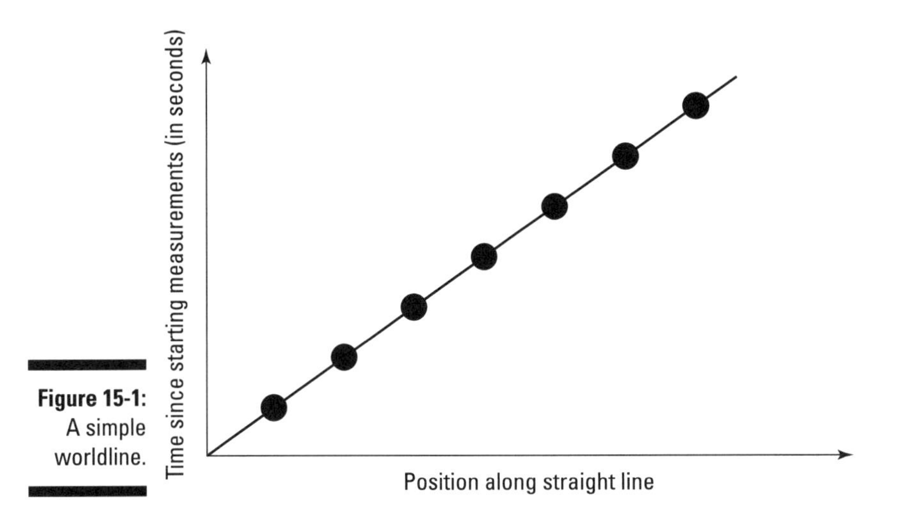

Figure 15-1: A simple worldline.

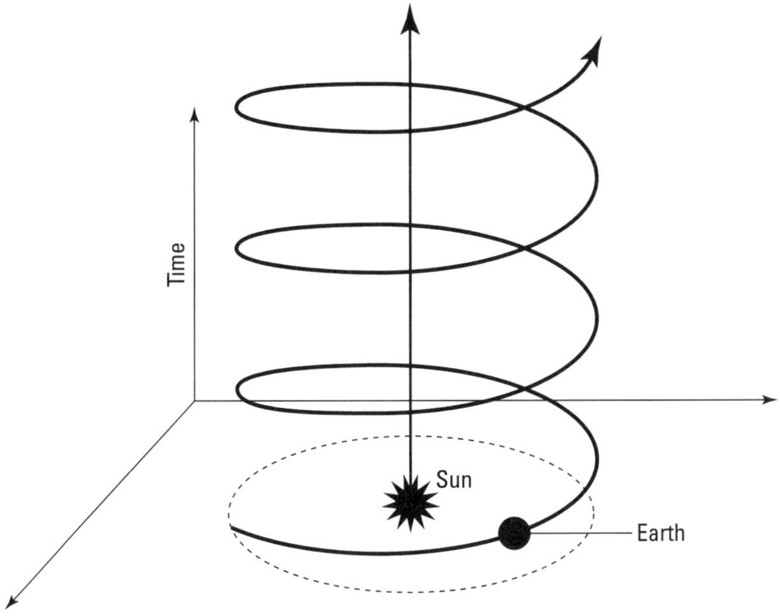

Time
Sun
Earth

Figure 15-2: The worldline of the Earth's orbit around the Sun.

The dotted line at the bottom of Figure 15-2 shows the elliptical orbit of the Earth around the Sun. Because you're showing how the Earth's position in space changes over time, you can plot it out as you did in Figure 15-1 and end up with the spiral that Figure 15-2 shows. Consider this drawing to be the worldline of the Earth in its orbit around the Sun.

The point of showing these diagrams is to get to grips with spatial dimensions. In Figure 15-1, you only have to contend with one spatial dimension (we say that the ball is constrained to move along that particular straight line). In Figure 15-2, we choose to ignore one of the space dimensions because we're looking at the Earth's orbit, which we say is flat for the purpose of simplifying things.

But real worldlines must take into account the three dimensions of space and the fourth dimension of time. Because we as authors are restricted to the two dimensional sheets of paper that make up the pages of this book, we can't make a helpful picture to show three dimensions, but you may now be able to imagine what an accurate depiction may involve. In fact, scientists often choose to ignore one of the spatial dimensions, just as we do with the Earth's orbit, when drawing worldlines.

The Sun turns green

What about the news that the Sun has turned green? How do you depict this in a diagram? Figure 15-3 shows the worldline of one photon of green light coming from the Sun (shown by the star). Again time is represented on the up and down axis. The bottom half of the diagram represents the past and the top half represents the future. In between these two is the moment when the photon is emitted by the Sun. Everything on that two-dimensional sheet that passes through the axes at the 'present' happens at the same time. When the photon is emitted, the Earth (the black circle) is shown at point A. The photon of green light (indicated by the black arrow) reaches the Earth after eight minutes, when it is at point B.

The Sun is also emitting green photons in every other direction; these photons are represented by the cone above the green Sun. Anywhere on the oval at the top of that cone can be receiving a photon at the same moment as you on the Earth.

The cone is known as a *lightcone*. The lightcone is usually drawn with 45 degree angle sides that project from the light source and includes all the particles that are travelling at the speed of light.

The lightcone concept is important because it represents the region of space in which an observer can possibly know about an event – in this case, the Sun turning green. As you can see in Figure 15-3, people on the Earth at point A are outside the lightcone. Their position means that they have no way of knowing that anything has happened to the Sun. At point B, they enter the lightcone and become aware of this strange celestial phenomenon.

Figure 15-3 also shows another lightcone below the green star. This area represents the time *prior* to the Sun turning green. Scientists can say that the event being represented – the emission of a green photon from the Sun – can be influenced only by events that are within the lower lightcone.

Anything outside the lower cone can not have had any influence on the event because it is too far away to have interacted with it.

Based on the lightcone concept, scientists define three types of paths within spacetime:

- **Lightlike** describes the paths of things travelling exactly at the speed of light, such as the green photons in the preceding example in which the Sun suddenly turns green.
- **Timelike** describes the path of things travelling at less than the speed of light – that is to say, any path that falls *within* the lightcone.
- **Spacelike** describes any path *outside* the lightcone – such as a line joining the Sun to the Earth at point A in Figure 15-3. Such a path involves travel (either of a piece of information or of an object) at a speed greater than that of light and as such is physically impossible.

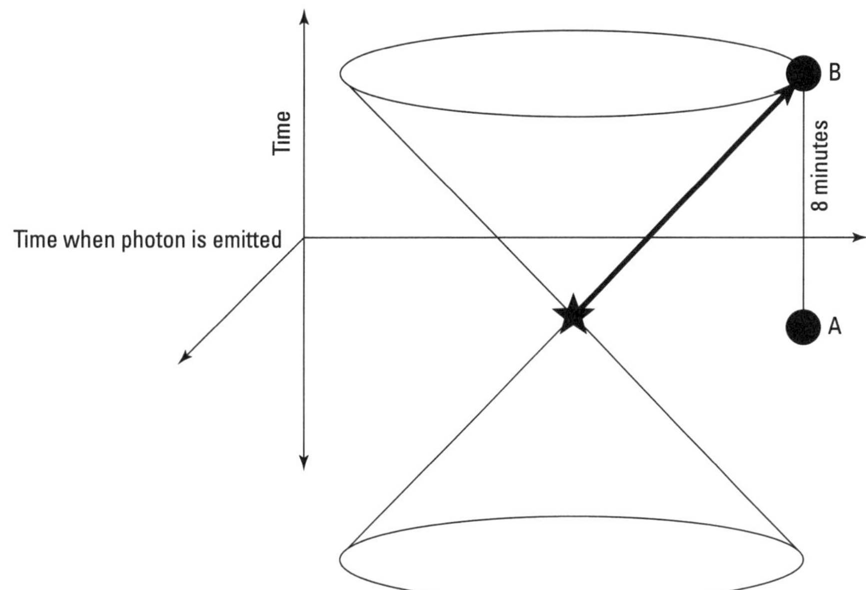

Figure 15-3: The worldline of a photon coming from the Sun.

The twins paradox

One of the strangest ideas to come out of Einstein's special relativity (which says that time and space are perceived differently depending on the movement of the observer, as we describe in Chapter 4) has been dubbed *the twins paradox*. The notion involves two identical twins – one who likes the home life and the other who's a bit of an adventurer. The first twin decides to stay at home on the Earth, while the second decides to zoom off on a ten-year round trip to the stars in a new spaceship that can travel at 60 per cent of the speed of light (which is still extremely fast). Both are wearing watches given to them by their mum and they synchronise them before the trip.

If Newton's ideas about the universe were correct (see Chapter 3 for more on Newton), the two watches would still be synchronised on the adventurer's return. However, Einstein theorised that the traveller's watch appreciably slows down because he's travelling so fast. As a result, when the travelling twin returns to the Earth, he's two years younger than his twin.

This scenario is called the twins paradox because it seems to violate a central point of Einstein's theory regarding *frames of reference* (which we discuss in Chapter 4).

You can think of frames of reference by considering a car crash involving two vehicles travelling towards each other at 100 kilometres per hour (60 miles per hour). A casual bystander sees the cars travelling at 100 kilometres per hour, but from opposite directions. For both of the people in the cars, it appears that the other car is approaching them at the combined velocity of 200 kilometres per hour (120 miles per hour).

Now consider things from the frame of reference of the travelling twin rather than the one at home on the Earth. In the travelling twin's frame of reference, the home twin appears to be whizzing away from him at 60 per cent of the speed of light. According to Einstein, it would seem to the travelling twin that it would be the home twin's watch that's going more slowly and that he (the stay-at-home sibling) would be the one who's two years younger.

The reality is that the travelling twin would be the one who ages more slowly. In fact, the paradox doesn't violate special relativity because you need to consider *three* frames of reference, not two. Specifically, you need to consider:

- The frame of the non-moving home twin
- The frame of the adventuring twin as he speeds away
- The frame of the adventuring twin as he zooms back

In Newtonian mechanics, the difference in velocity between the last two frames is 60 per cent + 60 per cent = 120 per cent of the speed of light – something that Einstein says is impossible. In fact, Einstein's way of adding velocities (the complexities of which we aren't going to go into here) means that the relative velocity of the last two frames is actually something like 88 per cent of the speed of light. For this reason, the twins paradox doesn't exist. Special relativity holds true and the adventuring twin always comes home younger.

The ideas encompassed by the concept of lightcones are important because they give you a way of handling what you see when you look at the universe. Because you can see events only where the light has had the opportunity to reach you, everything you can see is within the lightcone from the past. Your view of now is thus very much guided by what you can possibly have experienced.

Trawling for tachyons

Einstein's work and complex equations appear to show that the threshold of the speed of light cannot be passed. Objects with zero mass (such as photons) travel at that limit, but anything with mass – you or a spaceship, for example – can't reach that speed.

The problem is that as an object with mass travels closer to the speed of light its mass increases exponentially. Einstein's equations show that an infinite amount of energy is needed to accelerate an object with mass up to the speed of light.

But what scientists eventually realised is that although the speed of light represents an upper limit for the velocity of most things in the universe, it may be the lower speed limit for objects called *tachyons* (from the Greek for 'fast', just as in the word tachometer). Some theorists suggest that tachyons always travel faster than light and can never slow down to the speed of light or below. According to the theory, slowing tachyons down below the speed of light would take an infinite amount of energy.

Tachyons, which no scientist has ever observed, would have some very strange properties.

- **Additional energy would slow down tachyons.** If you gave a tachyon some extra energy, it would slow down – the exact opposite of what scientists experience with everyday particles (which they sometimes call *tardyons* or *bradyons*).
- **Tachyons would be very hard to see.** If a tachyon were coming towards you, you wouldn't see it until it had gone past.

However, if tachyons do exist, scientists should be able to detect the Cerenkov radiation (see the later section 'Travelling faster than light' for more on this phenomenon) emitted as they travel faster than the speed of light through another material. However, this type of radiation hasn't been observed from tachyons – leading most scientists to doubt their existence.

Considering causality

Causality is not the hospital department you would end up in if this hefty tome fell on your foot. Instead, *causality* is a description of the relationship between cause and effect – your dropping this book is the cause and your foot hurting is the effect.

Imagine how bizarre life would be if your foot suddenly hurt for no reason but then some time later you dropped a book onto your foot and you suddenly realised why it had been hurting.

If an alien in another part of the universe were looking through a very powerful telescope at you on the Earth, it would see events unfold in the same way you experience them – the book drops and then a while later a red bump appears on your foot.

That's what the alien would see if the speed of light really is a fundamental limit. However, if things can travel faster than light, the alien may be able to observe the bump on your foot *before* the book fell. This sequence of events is possible if tachyons exist. In some frames of reference, the absorption can be shown to happen before the emission – that is, the effect has happened before the cause. Simply put, if faster-than-light speeds are possible, causality is in big trouble.

Factoring in the speed of gravity

In recent centuries, scientists have begun to realise that limitations exist to causality. Take gravity, for example. Newton realised that a gravitational attraction exists between the Sun and the Earth, and he believed that the interaction was instantaneous.

However, Einstein's general theory of relativity showed that gravity is caused by the curvature of spacetime (Chapter 4 has more about Einstein's work) and that it travelled with a finite speed – the speed of light. Fluctuations in the curvature of spacetime, caused by a spinning neutron star for example, also generates gravitational waves, which travel at a fixed speed. And at what speed do gravitational waves travel? The speed of light.

To get an idea of these gravitational waves, imagine a watermelon wobbling around on a sheet of rubber. The wobbles cause ripples in the rubber, rather like the ripples on the surface of a pond.

This means that if the Sun were to disappear by some magical means, you wouldn't realise that it had gone for eight minutes the same amount of time

Is the speed of light slowing down?

A handful of scientists and a larger number of amateur cosmologists have wondered whether the speed of light itself is changing. If it were, the change would have profound implications for the age of the universe since the red-shifts we measured from distant galaxies would change over time, leading to a change in Hubble's constant (see Chapter 5) over time. We gain our estimates of the age of the universe from this constant. Some argue that if the speed of light has been dropping rapidly since the Big Bang, the universe may be only a few thousand years old (pleasing those who believe in creationism) rather than billions of years.

One of the leading proponents of this idea is Australian Barry Setterfield, who looked at historical measurements of the speed of light and argued that they showed that the speed of light had indeed fallen dramatically over time.

However, other theorists have shown that many other scientific constants are indirectly related to the speed of light, such as the charge on an electron and Planck's constant. If the speed of light has really been slowing down, the stellar spectra of very distant stars would be very different from those of nearby stars. But in fact, no such differences exist.

required for the light to reach the Earth. You would see the sunlight blink out and the Earth would go spinning off into deep space in a straight line rather than continue in its elliptical orbit.

However, if Newton had been right, the Earth would have spun off into space immediately. You would have wondered why it was happening because the Sun would still be shining – but only for another eight minutes.

Turning Back Time

Ever since H.G. Wells wrote *The Time Machine*, people have been fascinated by the idea of time travel – for example, the possibility of building a clever machine that takes you forward in time to find out the score in the next World Cup final and then travel back to the present to place a huge bet on the result.

In theory, travelling through time is just a matter of travelling faster than the speed of light. Although Einstein says that travelling faster than the speed of light is impossible, recent research indicates that his assumption isn't quite accurate, as we discuss in the following section.

Travelling faster than light

Faster-than-light travel isn't possible if you want everyday observations of cause and effect to be maintained. And yet, things can and do travel faster than the speed of light.

Cerenkov radiation

When Einstein said that things can't travel faster than the speed of light, he was referring to the speed of light in a vacuum. In fact, light can travel more slowly in different materials.

- In water, light travels at around three quarters of the speed of light in a vacuum.
- In glass, light travels at around two thirds of the speed of light in a vacuum.

So what happens when something travelling faster than light in one material enters another material in which the speed limit is slower? The answer is that it gives off energy in the form of radiation, *Cerenkov radiation* (named after the scientist who first explained the effect). You can observe Cerenkov radiation in a number of places. Nuclear reactors get their characteristic glow through Cerenkov radiation, and particle accelerators often feature detectors that look for Cerenkov radiation to chart the passage of a charged particle.

The expansion of space

Although Einstein stated that things can't travel in space faster than the speed of light in a vacuum, nothing is stopping space itself from expanding faster than this speed. In fact, if the inflationary model of the universe (which we discuss in Chapter 7) is to be believed, that is exactly what happened.

The age of the universe (the time since the Big Bang anyway) is thought to be 13.7 billion years, and yet research published in 2004 put the minimum size of the universe as 78 billion light years in diameter. If space were not being stretched, the expected size would be 27.4 billion light years (that is, double the universe's age because two light beams could be emitted in opposite directions at the time of the Big Bang).

Venturing Back to before the Big Bang

Chapter 6 covers the most commonly held theory of the origin of the universe – the Big Bang. The thing we don't really touch on in Chapter 6 is what caused the Big Bang to happen in the first place and what was around before it happened. (We talk at the beginning of Chapter 7 about what set the whole thing off, but we neatly sidestep what actually happened.)

Summing things up in a singular point in time

What happened before the Big Bang is one of the hardest concepts relating to the origin of the universe to grasp. The problem is that many simplified models of those earliest moments lead people to assumptions that may not be justified.

For example, many people talk about the Big Bang as an explosion that caused the universe's expansion. But the Big Bang wasn't an explosion in the commonly understood sense.

Most theories of the origin of the universe start off with a *singularity*, a starting point at which both the density of matter and the curvature of spacetime are infinite. You can envision this starting point by running the universe's clock backwards – at least in your imagination.

Hubble's Law (turn to Chapter 5 for more on Hubble) shows that galaxies are moving away from the Earth at a velocity proportional to their distance away from the Earth. At some time in the distant past, all these galaxies and indeed every other bit of matter in the universe was crushed into a single point. In other words, the density of the entire universe at that point was infinite.

At a singularity, all the laws of physics break down. You can see something similar to that mathematically with the reciprocal function $1/x$, which Figure 15-4 shows in graph form.

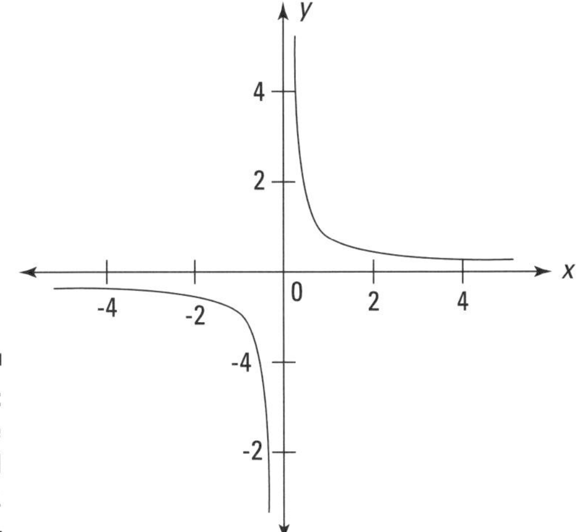

Figure 15-4: The reciprocal function $1/x$.

If you start with a positive value of x and move closer to zero, $1/x$ gets larger and larger. If you start with a negative value of x and move closer to zero, $1/x$ gets more and more negative. You can see that something weird must happen when x is exactly zero because the value of $1/x$ must suddenly jump from negative infinity to positive infinity. Mathematically speaking, the value of the reciprocal function $1/x$ is *undefined* at $x = 0$, and you get a discontinuity. (For this reason you get an error if you tap in zero on a calculator and then press the $1/x$ button.)

Consider the Big Bang to be that point where x equals zero. Whatever happened at that point, a discontinuity existed. Just like the $1/x$ function, the density rose to infinity at that point. If something did exist 'before' the Big Bang, there would have been a discontinuity in any law of physics you care to consider at the point of creation.

The upshot of this discontinuity is that no physical observation anyone can make now can provide any insight into what happened before the Big Bang – if indeed anything even existed before that point. Put another way, time as you know it began with the Big Bang.

In fact, you can go a step farther: One of the most important things to remember from the work of Albert Einstein and Hermann Minkowski, a Lithuanian-born mathematician who was one of Einstein's teachers, is that space and time are inextricably linked in the concept of spacetime. The Big Bang was thus not just the point at which time was created – but when space was created too.

Contemplating an infinite universe

One obvious get-out clause for anyone worrying over what happened before the Big Bang is to assume that the universe has been around forever. But this idea also has problems.

Why? Well, the story begins back in the 19th century. One of the great advances of scientists of that era was in the understanding of *thermodynamics*, or changes in temperature. In the mid 1800s, German physicist Rudolf Clausius set down what has become known as the *second law of thermodynamics*. In the simplest terms, the law says that differences in things such as temperature and pressure tend to even themselves out over time. Not such an odd notion, right? Think how a hot glass and a bowl of cold water eventually both settle at some temperature in between the two when one is placed in the other.

But if the universe really is infinitely old, any differences in temperature would have disappeared long ago, and hot things such as stars wouldn't be lying around the place.

In Chapter 5, we also talk about Olbers' Paradox, which has to do with the fact that the night sky is dark, when in an infinite universe the sky should be filled with the light from infinite stars. The fact that the entire sky isn't ablaze with starlight is another good reason for thinking that the universe isn't infinitely old.

Pondering the turtle problem

The problem of 'what happened before?' isn't restricted to the Big Bang and other scientific theories about the origins of the universe.

If you believe that God created the universe, you can quite reasonably ask what existed before he did so. You may even be tempted to wonder about the nature of God. If he did create the universe, has he been around for an infinitely long time? Or perhaps someone created him? And if that someone created him, who created that someone?

These infinite regressions can be classified under the heading of 'turtle problems'. The name refers to various stories (check out Chapter 19 for more on alternative theories of the creation of the universe) in which the Earth or parts of it are standing on the back of a giant turtle. A cynic may ask: 'What is the turtle standing on?' to which you may answer 'Another turtle'. The cynic would then reply 'But what is that turtle standing on?' The only way to avoid getting into a never-ending philosophical argument is for you to reply, 'It's turtles all the way down.'

Part IV

Asking the Tough Questions

'If you'd been baked one minute, frozen stiff the next, bombarded with radiation, choking in endless duststorms, continuously gasping for breath in a thin atmosphere, *you'd* have a bad attitude too!'

In this part . . .

In this part we try to answer the difficult questions about the universe. What are black holes, white holes, and wormholes? Are parallel universes simply science fiction? Are we alone in the universe? Read this part to find out.

This book is about the origins of the universe, but it's hard to avoid the tough questions about how the universe is going to end. Will it be a Big Chill, a Big Crunch, or a Big Rip? However the cosmos comes to a close, it's almost certainly going to be big. It's the universe after all.

Chapter 16

Explaining the Unexplainable

In This Chapter

- Examining some of the wildest stuff in the universe
- Attending a star's funeral
- Escaping from a black hole
- Explaining pulsars and quasars
- Pondering parallel universes

Have you ever wondered where all your socks disappear to? You know how it goes – every time you do a load of washing, you carefully pop nice pairs of socks into the machine, two by two. But no matter how diligent you are, they don't stay together. Before long you find yourself sitting on the bus, wearing pink on one foot and green on the other. Somewhere, somehow, the others have vanished.

What's to blame? A black hole? A wormhole? A rip in the fabric of spacetime that transports your footwear to a parallel universe where wild socks roam free on the prairie?

Who knows. But one thing's for sure: Disappearing socks shouldn't be underestimated as one of the inexplicable phenomena of everyday life. You can add them to a list of strange stuff going on in the universe – including black holes, neutron stars, quasars, and so on – which is our focus in this chapter.

Watching Stars Die

Stars are the result of clouds of gas and dust collapsing in on themselves under gravity, as we describe in Chapter 12. As stars form, the hydrogen atoms in the gas cloud collide with one another more and more frequently, with greater speed and energy, until they eventually start fusing together to make helium. The heat released by these interactions makes stars shine with the brilliant light you see in the sky day (in the case of the Sun) and night.

Depending on the size of a star and the amount of hydrogen fuel it starts off with, a star can burn with this glorious atomic energy for millions or even billions of years. During this phase of its life, the inward pull of its gravity is balanced by the outward push of pressure from all the molecules packed inside it.

But one day, this beautiful balance comes to an end. The star's core runs out of fuel. The inner region of the star grows ever more compact, collapsing under gravity, while its outer regions, which still contain material undergoing nuclear fusion, start to expand outward, forming what's known as a *red giant*.

For some stars, this process ends in a kind of petering-out, and they end their days as slowly cooling white dwarfs. This fate is eventually going to befall the Earth's Sun, another 5 billion years down the track.

Bigger stars, however, live faster and die more spectacularly. They're the James Deans of the universe, going out in dramatic fashion, exploding into cataclysmic supernovae. Some such stars end up as *neutron stars*, a remarkable creature we describe in the later section 'Knowing Neutron Stars'.

For the most massive stars of all, the implosion of their cores stops for no one, reaching infinite density and forming what scientists call a *singularity* (see Chapter 15 for more on them). The gravitational attraction becomes so strong that nothing can escape – and the result is a black hole.

Being Aware of Black Holes

Black holes are among the most fascinating things in the universe – and not just because they're the most likely culprit in the disappearing sock mystery.

Modern scientists define *black holes* as regions of spacetime (refer to Chapter 15 for more on spacetime) where the effect of gravity is so strong that nothing can escape, not even light. You can fall into a black hole, but you can't climb out.

Creating black holes

An English clergyman named John Mitchell first proposed the idea of black holes back in 1783. The more recent history of black hole science begins with one of the earliest solutions of the equations of Albert Einstein's general theory of relativity, which we describe in Chapter 4.

That solution, worked out by Karl Schwarzschild in 1916, was originally intended to describe the mathematical basis for stars, but scientists soon realised that any object can form a black hole if it shrinks below a certain size (called the *Schwarzschild radius*).

To calculate the Schwarzschild radius you use a simple formula:

$$r_s = 2Gm/c^2$$

To explain, the formula means that for an object to form a black hole, it needs to shrink to a size that equals two times the gravitational constant (refer to Chapter 3), multiplied by the mass of the object in question, all divided by the speed of light squared.

In terms that are a little more concrete, this formula means that for an object with the mass of the Earth to form a black hole, the object needs to be squished inside a radius of just 1 centimetre. For the Earth's Sun, the crucial radius is about 3 kilometres.

In terms of relativity, black holes warp spacetime just as any object with mass does. (Remember our idea of a watermelon on a sheet of rubber to explain how massive objects warp spacetime in Chapter 4). But because black holes are so dense, they have a more radical impact than other objects. Take a look at Figure 16-1 to get a sense of how an object's impact on spacetime changes as the object gets more and more compact.

Each of the balls on the left have the same mass (but different densities). The smaller and more dense the ball, the more it warps spacetime.

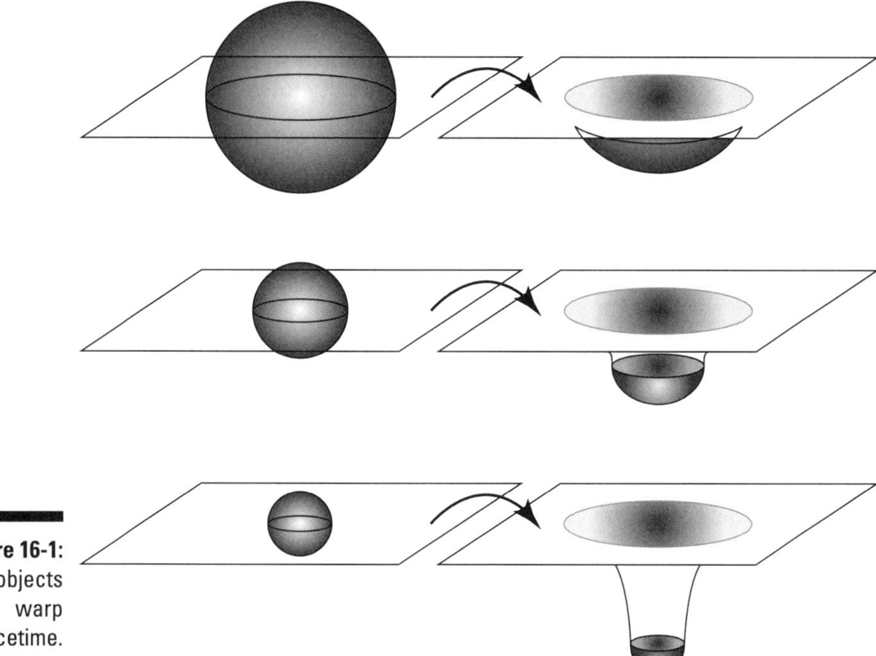

Figure 16-1: How objects warp spacetime.

Seeing the invisible

If a black hole doesn't allow any light to escape from it, how can astronomers find one – other than by tripping over it in the dark?

Scientists have a few indirect methods they use to look for black holes:

- **Accretion discs.** Clues from the gas being drawn in by the gravity of the black hole can suggest the position of black holes. These swirling *accretion discs* form around other objects too, but they can be good starting points for astronomers hunting for black holes. Supermassive black holes at the centre of galaxies are thought to cause spectacular accretion discs.
- **X-ray emissions.** As gas falling into the black hole forms an accretion disc, the gas gets hotter and denser, shining and emitting radiation in the form of X-rays and visible light.
- **Objects orbiting black holes.** Identifying objects that orbit black holes can help scientists spot black holes. Some scientists believe, for example, that the Milky Way has a supermassive black hole at its heart. By watching how stars orbit whatever is at the centre of our galaxy, scientists can work out how massive it is.

Categorising black holes

You can see from the Schwarzschild formula in the earlier section 'Creating black holes' that black holes can theoretically form in a whole range of different sizes. All that matters is that enough mass is packed into a small enough volume.

In reality, though, some objects aren't likely to form black holes. So scientists classify back holes into a discrete list of types, based on their size and how they formed.

- **Supermassive black holes** are thought to contain perhaps billions of times the mass of the Earth's Sun and exist in the centre of most galaxies, including the Milky Way.
- **Intermediate-mass black holes** are black holes whose size is measured in thousands of solar masses and only a handful of them have ever been discovered. Those that have been observed appear to reside in star-forming areas of the universe, although how they form is still a mystery.
- **Stellar-mass black holes** are roughly the mass of a large star and are created by the collapse of individual stars. Scientists recognise them by the strong X-rays they omit. Several objects within the Milky Way, including the X-ray source Cygnus X1, are believed to be stellar-mass black holes. Given their role in stellar evolution, these black holes are likely to be widespread throughout the universe.

- **Mini black holes** are (what else?) tiny black holes that scientists think may have been created during the Big Bang. Objects with mass of less than three times that of the Sun aren't normally big enough to collapse into black holes. However, the high pressures in the moments after the Big Bang may have been sufficient to make it happen. These mini black holes might be smaller than an atom but might have a mass of a trillion kilograms.

Looking inside black holes

Black holes consist of two main parts:

- The **event horizon,** which is a spherical surface that defines the edge of the black hole.
- The **singularity,** which is the heart of the black hole, where matter is infinitely compressed.

The event horizon is the defining feature of a black hole. It's the limit of the area within which light, or anything else, has no hope of getting out again (at least not without the help of something called Hawking radiation which we describe in the sidebar 'Are black holes really so black?'). Because nothing can escape, after a black hole has formed it remains a black hole forever.

If you were looking directly at a black hole in space, with something nice and bright behind it, you'd see the event horizon as a black sphere. However, in reality, black holes are usually obscured by brightly glowing dust, gas, and stars that are falling into them.

Within the shroud of the event horizon, things are a bit of a mystery. Using general relativity, scientists calculate that the heart of the black hole is a singularity. *Singularities* are points of perhaps infinite density, which have no dimensions – no length, width, or height.

Singularities defy current human abilities to describe them accurately. But we can safely say that space and time as humans know them are pretty messed up in there.

Falling into a black hole

Despite what you may have seen at the movies, black holes don't drag matter into them. It is perfectly possible for other objects outside the event horizon to orbit black holes without being sucked in like water down the drain. On the other hand, getting yourself mixed up with a black hole isn't likely to be pleasant.

Spaghettification

Because the effect of gravity becomes weaker the farther away you are from the object generating the field (thanks to the inverse square law we discuss in Chapter 3), scientists expect that anyone or anything getting close to a black hole experiences something called a *tidal force*, stretching them lengthways in the direction of the black hole.

The strength of the tidal force depends on the size of the black hole. Smaller black holes are thought to exert stronger forces, whereas enormous supermassive ones such as those thought to be at the centre of galaxies exert milder effects.

Imagine that you want to calculate the tidal forces on something 1 metre tall, such as a small child, located at the event horizon of a black hole with a radius of 1 metre. The child's toes are 1 metre away and the head 2 metres away from the centre of the black hole. Because gravity obeys an inverse square law, the force on the head will be four (two multiplied by two) times weaker than the force on the toes – or, to put it the other way round, the force on the toes is four times stronger.

However, if the child is at the event horizon of a 100-metre radius black hole, the difference between the forces is a lot less – the difference between 100×100 and 101×101, or just over 2 per cent, in fact. This means that the child's body would be pulled apart by tidal forces much faster in the smaller black hole than the larger one. (Yuk.)

In smaller black holes, scientists think that the differences between the effects of gravity on your feet and head would 'spaghettify' you – stretch you long and thin like a piece of spaghetti, perhaps before you even cross the event horizon.

Messing with time

Another effect of falling into a black hole is the strange behaviour of time.

Imagine, for example, that a brave robot called Marvin has volunteered to enter a black hole in a small spacecraft. On the outside of the ship, a great big clock displays the time as measured by an on-board clock.

Also imagine that you're sitting on another ship some distance away watching Marvin's journey through a telescope. As he gets closer and closer to the black hole, the time on the side of the ship runs more and more slowly (in accordance with Einstein's view that time slows down in a massive gravitational field) until it stops. To you, Marvin's ship would seem suspended just before the event horizon. This view wouldn't last forever. Because the light from Marvin's ship is travelling in the form of photons, at some point the last photon emitted before the ship entered the event horizon will reach you and Marvin's ship will just fade away.

Are black holes really so black?

The British physicist Stephen Hawking made one of the most amazing discoveries about black holes in 1974. Hawking showed that something *can* escape from black holes, in theory at least. Hawking showed that black holes could emit *thermal radiation*, or heat. In doing so, he showed that black holes aren't as black as scientists previously supposed.

How so? Well, Hawking's calculations are based on a fascinating aspect of quantum physics known as *virtual particles*. According to quantum physics, empty space isn't empty at all, but is filled with particles flashing into existence then blinking out again after a tiny fraction of a second. Virtual particles like these always come in pairs, one particle and one anti-particle, such as an electron and a positron.

The whole of space is filled with these virtual particles, which normally quickly annihilate each other. But near the event horizon of a black hole, one half of the pair can fall into the hole, while the other escapes. The resulting particles are called *Hawking radiation*. Although the theory sounds plausible, this radiation has never been observed . . . yet.

For Marvin, on the other hand, the experience is quite different. He passes over the event horizon with no trouble (other than being ripped apart by tidal forces and perhaps fried by other aspects of black holes). After he's inside the event horizon, he wouldn't be able to see anything outside the black hole's event horizon any more and never will again.

Breaking through to the other side

What if black holes aren't a dead-end but lead somewhere else? This hypothesis is beloved of science fiction writers – and the occasional scientist.

Black holes with exits on the other side are known as *wormholes*, and although no evidence suggests that they exist, they're interesting to think about.

The name wormhole gives a pretty accurate description of how these hypothetical phenomena may work. Just as an ant crawling over an apple can take a short-cut through a hole that has been munched by some wormlike critter, matter can do the same thing in space, or even time. (If such things as wormholes exist, that is!)

Some scientists suggest that certain black holes may actually lead to their opposite – a 'white hole' – where matter and light pour out. Others imagine black holes serving as gateways to other universes. We talk more about the possibility of other universes in the later section 'Creating Parallel Universes'. For the meantime, however, we concentrate in the following sections on a few more fascinating things that really do exist.

Knowing Neutron Stars

When the biggest stars in the universe (and no, we don't mean Madonna) reach their dramatic finale, one possible outcome is the formation of a black hole. Another possible result from the collapse of slightly less gigantic stars is the formation of a cold, enormously dense neutron star.

Neutron stars (see Figure 16-2) are thought to result from the deaths of stars that are between 9 and 30 times the size of the Earth's Sun. After billions of years of turning hydrogen into helium and progressively heavier elements, these stars begin to accumulate a core of iron.

Figure 16-2: A neutron star (shown by the arrow) as seen by the Hubble Space Telescope.

Courtesy of Fred Walter (State University of New York at Stony Brook) and NASA

Up to a certain point, this core is able to resist the inward pressure of gravity, but after the core reaches about 1.44 times the mass of the Earth's Sun (called the Chandrasekhar mass), the star gives up the ghost and collapses inward in as little as a second.

When this collapse happens, all the protons and electrons in the core of the star are squashed so close together that they interact to form neutrons and neutrinos, the latter of which carry off energy.

Eventually, degeneracy pressure stops the shrinkage, and the rest of the star's matter bounces off the neutron core like a wave hitting a sea-wall, triggering a cataclysmic explosion that scatters the rest of the star far into space and shines as bright as all the stars in a distant galaxy. Boom. That's a supernova for you (refer to Chapter 12 for some super supernova facts).

What's left behind is some of the densest material in the universe, a neutron star. Neutron stars are roughly 10 kilometres across but weigh more than the Sun. That's 100 trillion grams per cubic centimetre. A teaspoon full of this stuff would weigh a billion tons on the Earth, so you wouldn't want to mistake it for sugar and try to pop it into your tea.

Pondering the Pauli exclusion principle

So what keeps neutron stars from shrinking any further? The force at work is called the *Pauli exclusion principle*, one of the key rules of quantum mechanics.

In fact, the Pauli principle, which was formulated by Wolfgang Pauli in 1925, explains why all matter occupies its own space and doesn't allow other objects to pass through.

In quantum jargon, the rule says that no two identical fermions may occupy the same quantum state simultaneously. A *fermion* is the name scientists give to particles, such as a neutrons or electrons, which possess certain values of a quantum property known as spin (see Chapter 10 for more on this).

The Pauli exclusion principle explains why you can't walk through walls, and why tennis balls bounce off rackets rather than flying straight through them. The principle also explains why neutron stars don't get any denser – their neutrons are packed together as tight as can be.

Checking the pulse of neutron stars

Some neutron stars go by a different name. They're called *pulsars* because they emit a beam of radio waves, X-rays, gamma rays, and/or visible light that shines across space like the light from a lighthouse.

Spin, neutron star, spin

Neutron stars start their lives spinning like crazy – hundreds of revolutions per second. If you've ever watched ice skating, you can understand the reason why.

If you think back to the last Winter Olympics you saw, you may remember how skaters often start a spin with their arms stretched out. As they bring their limbs in closer to their bodies, their spins get faster and faster.

Physicists explain this cool trick by the term *conservation of angular momentum*: It's impressive on ice, but mind blowing when it comes to stars. When compact stars form from the inward collapse of bigger stars, their rotation accelerates enormously.

Scientists think that the beam of radiation pulsars emit results from the fact that the star's magnetic axis isn't lined up with its rotation. As the star spins, that beam sweeps across your line of sight here on the Earth, which your telescope or detector picks up as a regular pulse.

Meeting Quasars, the Fascinating Hearts of Galaxies

In the 1950s, scientists using radio telescopes started finding objects out in space that emitted strong radio waves and looked like stars when viewed through ordinary visible light telescopes. They called these things quasi-stellar radio sources, or *quasars* for short.

Since then, astronomers have detected many more similar objects, most of which don't actually emit strong radio waves, and so they tried changing the name to quasi-stellar objects (QSOs), but the name quasar is catchier, so it stuck.

As scientists studied these mysterious objects more, they found that many were highly red-shifted. *Red-shifting* is an effect that results from the expansion of the universe. As we describe in Chapter 5, the farther away an object is, the more its radiation is red-shifted.

Light coming from so far away has also taken an awfully long time to reach you here on the Earth, which suggests that the light from some quasars has taken more than 90 per cent of the age of the universe to reach the Earth. Looking at quasars, therefore, is like looking back through the history of the universe.

So what are quasars? Well, one early suggestion was that quasars are 'white holes', a kind of reverse black hole that we talk about in the earlier section 'Breaking through to the other side'. But scientists currently don't think that this explanation is true.

One of the best clues about the nature of quasars is the fact that their emissions vary over years, months, weeks, and even over the course of a single day. That's a sure sign that quasars are pretty small in cosmic terms (because light must be able to travel from one side of the object to the other in that same timescale).

The varying energy emissions of quasars, along with other details, has led scientists to conclude that quasars are powered by matter falling into giant black holes at the centres of distant young galaxies.

Quasars are an extreme form of something that astronomers call *active galactic nuclei* – a bright central region in a galaxy caused by the accretion of material into an enormous black hole.

Creating Parallel Universes

As we mention in the earlier section 'Being Aware of Black Holes', all this talk of black holes, white holes, and quasars starts a lot of people thinking about *parallel universes*, a tricky term we define in the following section.

You may think that such talk is nothing more than feckless science fiction mumbo-jumbo, but in fact parallel universes are a topic of genuine discussion among cosmologists. The following sections eavesdrop on some of these debates.

Taking a trip through infinite space

Before we start talking about parallel universes, we need to pin down what we mean by the word *universe*. For practical purposes, the universe is the patch of space that you can possibly observe. This patch is technically called the *observable universe*, and it stretches out as far as light has been able to travel since the Big Bang roughly 14 billion years ago.

The most distant visible objects are now about 4×10^{26} metres away (that's 4, followed by 26 zeros). A sphere of that radius makes up the observable universe.

What's beyond that boundary? More space? The answer depends on the overall curvature of space. A few different options for the larger universe exist. The most recent evidence, based on an analysis of the cosmic microwave background (refer to Chapter 6 for more on this relic radiation of the Big Bang) suggests that the universe is likely pretty flat. If that's the case, the universe may also be infinite in size and teeming with galaxies, stars, and planets.

If the preceding scenario is correct, an enormous number of other patches of space like ours exist out there. In a scientific paper published a few years ago, the cosmologist Max Tegmark did some calculations based on this idea. According to Tegmark's work, if you assume that the universe is infinite and matter is evenly distributed throughout it:

- About 10 to the 10^{29} metres away from here, a closely identical copy of you exists.
- About 10 to the 10^{91} metres away, a sphere with a radius of 100 light years exists, which is identical to the one around you currently. So everything you perceive for the next 100 years is also the same over there.
- About 10 to the 10^{115} metres away, an entire other universe exists that's identical to the one you're part of right now.

So even though the universe is pretty diverse, in an infinite universe with stuff evenly distributed in it, you'll eventually come across a copy of yourself. This is spooky stuff. Only a few cosmologists really believe in this view of the universe.

Making bubbles with inflation

Another potential way to make parallel universes is via a process cosmologists know as *inflation*. We're not talking about the kind of inflation that makes the price of bread go up, but cosmic inflation – the idea that in the first moments of its history, the universe went through a short period of enormous, accelerating expansion. We describe inflation in more detail in Chapter 7.

The general concept of inflation answers several puzzles about the overall structure of the universe and has become widely accepted among scientists. The details, however, are up for debate. Over the years, cosmologists have developed various different models to describe how inflation may have worked.

One of these models, proposed by US-based physicist Andrei Linde, describes a variation known as *chaotic inflation*. In Linde's version, tiny quantum fluctuations in the fabric of the universe result in a host of inflating bubbles of space-time, each of which in turn spawns its own bubbles, and so on, forever and ever. The result? A universe divided into countless different unconnected 'buds' – in essence, a *multiverse* of different universes, each with its own laws of physics.

Chapter 17

Finding Life Elsewhere

In This Chapter

- Searching for life on Mars and Titan
- Finding planets outside the Earth's solar system
- Questioning the likelihood of alien civilisations
- Listening and looking for signs of intelligent life

Are humans alone in the universe? Did life emerge just once in the vastness of space – and as such is the Earth just an enormous (you may say cosmic) fluke?

Put another way: Of all the billions of planets in billions of galaxies in the observable universe, is the Earth unique in playing host to an intelligent form of life? Or are 'they' out there somewhere, other forms of life perhaps closer than we think – aliens!?

You need only to browse a small selection of the countless alien-infested novels, movies, or Web sites to understand how often humans ask these questions, even if just for fun. The questions are, after all, some of the most provocative philosophical issues humanity has grappled with. They're also questions that have no definitive answer – not yet, anyway.

And yet the search for life, or potential life, beyond the Earth has gained credibility in the past decade. Developments in chemistry, geology, astronomy, planetary science, oceanography, physics, and biology have come together in a new field called *astrobiology* – the scientific study of the living universe – which has the backing of serious agencies such as the National Aeronautics and Space Administration (NASA).

The development of astrobiology means that NASA and the European Space Agency (ESA) have joined groups such as the Search for Extraterrestrial Intelligence (or SETI) Institute in probing the universe for signs of life. Read on to find out what they're discovering about life beyond the boundaries of planet Earth.

Searching for Life in Our Solar System

Where to begin searching for life beyond the Earth? Well, where better than in your local neighbourhood – the planets and moons in your very own solar system. After all, they're the only places in the universe that humans (or human-controlled probes, actually) are currently equipped to get to!

Mars

Aliens from Mars are the ultimate sci-fi cliché, but it turns out that within the solar system, the red planet is arguably the most likely to harbour life. Of course, we're not talking about thin green beings with big heads and slanty eyes. The life, if it is there at all, is more likely to be microscopic life forms, perhaps bacteria.

Several reasons lead scientists to think that life may exist, or may once have existed, on Mars.

Finding methane

Scientists showed in 2003 and 2004 that the atmosphere of Mars contains methane. On the Earth, living things, such as the bugs that live in the stomachs of cows, produce the vast majority of methane.

The most likely explanations for the presence of methane on Mars are:

- **Methane-producing bacteria.** On the Earth, a group of micro-organisms called *methanogens* create the gas as a by-product of consuming hydrogen and carbon dioxide or carbon monoxide. Most methanogens don't need oxygen to survive and they've been found in some of the harshest environments on the Earth, suggesting that they could survive on Mars too.
- **Serpentisation.** Serpentisation is a chemical reaction between rock and water that takes place on the Earth – and perhaps on Mars. In the extreme conditions of deep sea hydrothermal vents, certain rocks can react in a way that produces hydrogen. Methane is formed when that hydrogen combines with carbon, carbon dioxide, or carbon monoxide.

Mars exploration is still in its early stages – humans have never been there and robotic exploration is in its infancy – so the answer to whether life exists on Mars may be some time in coming.

Seeing water flow

Water is essential for life as humans know it. The surface of Mars shows plenty of signs that liquid water was once abundant on the red planet. Evidence indicates the presence of ice on and below the surface.

Further evidence suggests that water sometimes still flows on the Martian surface. In late 2006, researchers examining evidence gathered by a NASA project called the Mars Global Surveyor reported something remarkable. As shown in Figure 17-1, they found two gullies on the surface of the planet that showed distinct evidence that water had flowed down them some time in the past five years.

The camera on board the Surveyor snapped before and after photos of two gullies in 2001 and 2005. The images from 2005 show fresh, light deposits of the kind that are formed when water flows down a channel.

The robotic Mars Explorer mission also found evidence of water vapour in the Martian atmosphere, and Martian ice caps are primarily made of water ice. Mars has no shortage of water – it's just liquid water that seems to be in short supply.

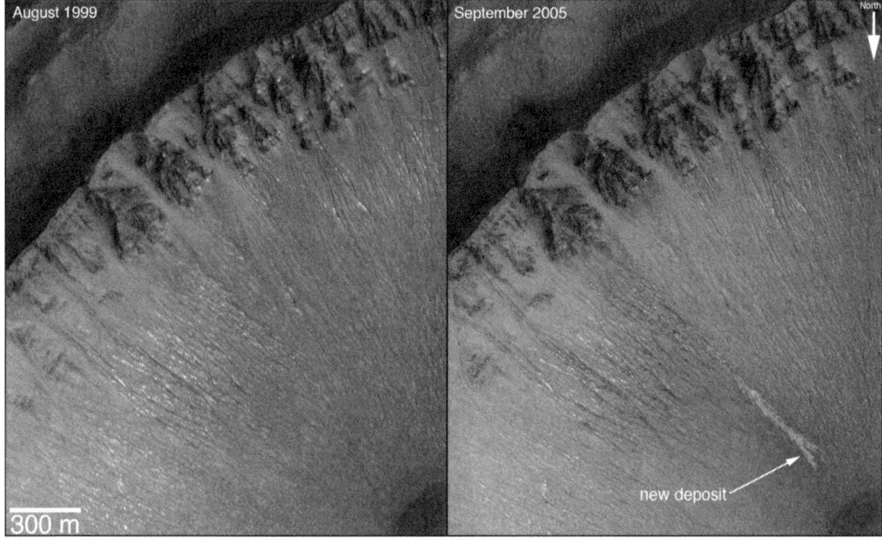

Figure 17-1: Before and after images showing that water flowed briefly on the surface of Mars in recent years.

NASA/JPL/Malin Space Science Systems

Titan

Titan, the largest moon of Saturn, is another place that people often talk about as a possible location for life beyond the Earth. Although Saturn is far from the Sun, meaning that it's much colder than the Earth, its thick atmosphere is rich in organic compounds, some of which would be considered signs of life if they were on the Earth.

In fact, Titan's atmosphere, like the Earth's, consists mostly of nitrogen, with methane and organic compounds added for good measure. Figure 17-2 shows a recent image of Titan, showing a bright patch of clouds that may consist of a mix of methane and ethane.

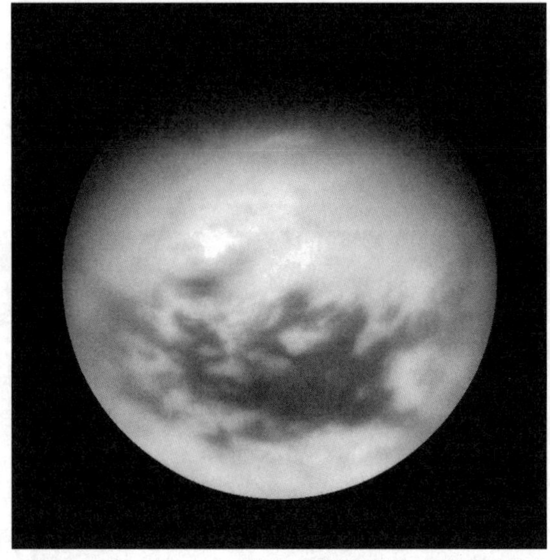

NASA/JPL/Space Science Institute

Figure 17-2: A photo of Saturn's moon Titan taken by the Cassini spacecraft in 2007.

Interestingly, organic compounds form when sunlight breaks down methane, so the question arises – if sunlight is continuously destroying methane, how is methane getting into Titan's atmosphere? Can life be forming it? Scientists think that it's unlikely. For one thing, Titan's surface is –180 degrees Celsius, far too cold for water in liquid form to exist.

Making a home on Mars?

As things stand, Mars doesn't strike anyone as a good place for humans to set up home. The planet is freezing cold, with a very sparse carbon dioxide atmosphere and hardly any running water. Billions of years ago, things were likely much more hospitable – and in the minds of some scientists, they may be again in the future.

The idea is not that the planet is going to become more welcoming of its own accord. Instead, scientists think that it may be possible to give the planet an extreme makeover – providing it with an atmosphere, turning its craters into lakes, and covering its barren hills with trees – so that it better suits human needs. This theoretical concept, known as *terraforming*, is another one of the many ideas that were once the realm of fiction writers, but are now popping up in respectable scientific circles.

Turning Mars into a planet that we can live on in comfort would take decades and enormous amounts of money – if it can be achieved at all. A crucial step is finding a way to enrich its atmosphere and warm the planet. If the planet's polar ice caps can be melted, releasing carbon dioxide, both goals could be achieved with one stroke. In other words, humans would trigger global warming on Mars.

Scientists think that this type of warming may be technically feasible – perhaps giant mirrors can melt the ice caps – but the process may not be desirable. At a time when the possibility of finding alien life forms on Mars seems closer than ever, terraforming may destroy such life at a stroke.

Finding Planets with Life Outside Our Solar System

Looking beyond the Earth's solar system, one of the first steps in searching for possible sites of alien life is to find planets on which those life forms may exist.

Astronomers presume that plenty of other planets exist in the universe because the birth of a star inevitably leaves behind the kind of messy residue that can easily accumulate into orbiting planets (refer to Chapter 13 for how planets are formed). Finding these planets isn't as easy as talking about them, though.

And yet finding other planets *is* possible, if you're clever enough. Scientists use a few main methods for detecting distant planets, which rely on spotting the impact that the planets have on the star they orbit, instead of seeing the planet directly. Technically speaking, these methods include:

- Measuring the radial velocity of a star
- Using astrometry to watch stars move
- Watching planets transit across stars

We describe and explore each of these methods in detail in the following sections.

Seeing starlight wobble: Radial velocity

Since scientists made the first definitive detection of an extra-solar planet (one that orbits a star other than the Earth's Sun) in the mid-1990s, more than 240 planets have been identified, and the number is growing quickly.

The most successful method for detecting planets so far has been the *radial velocity method* (also sometimes called Doppler spectroscopy), which involves looking for subtle changes in the spectrum of light coming from a distant star.

With the radial velocity method, astronomers look for clues in the star's light to indicate that a planet orbiting the star is affecting its movement, as illustrated in Figure 17-3. This technique is clever stuff. As a planet orbits the star, the planet's gravitational field pulls on the star, so that sometimes the star is moving toward, or away from, the Earth.

When the star is moving towards the Earth, the wavelengths of the spectral lines in the light it emits move towards the blue end of the spectrum, and when the star travels away, the wavelengths are moved towards the red part of the spectrum. (This effect is the same as the Doppler effect, which makes police sirens sound higher in pitch as they approach you, and lower as they drive away; refer to Chapter 5 for more on the Doppler effect).

Astronomers can use this variation in wavelength to look for stars where the spectral lines are moving back and forth – because these stars must be the ones with planets in orbit around them.

Related to the radial velocity method is a technique called *pulsar timing*, which looks at variations in the timing of the pulses of radiation emitted by a type of neutron star called a pulsar. This method was used to detect an extra-solar planet in 1994. (For more on pulsars, check out Chapter 16.)

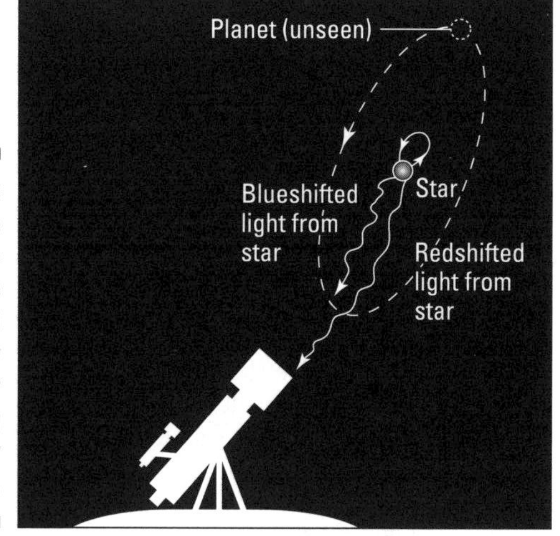

Figure 17-3: The principle behind the radial velocity method for finding extra-solar planets.

Doing the planet waltz: Astrometry

Instead of detecting the impact of the planet on a star's light (as in the radial velocity method), *astrometry* looks at the actual movement of the star in the sky, as we show in Figure 17-4.

Attempting this method using ground-based telescopes hasn't been particularly successful at detecting planets, because the changes in the star's position are just too small for even the best telescopes to detect. But scientists have had more luck using the orbiting Hubble Space Telescope, which isn't impeded by the Earth's atmosphere. In this way, astrometry has detected a small number of extra-solar planets circling nearby stars.

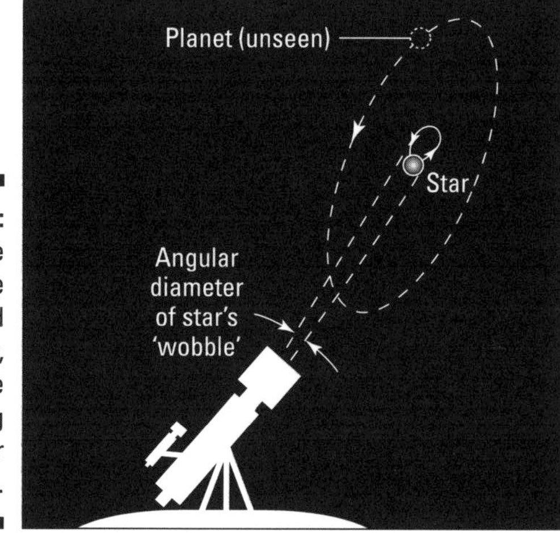

Figure 17-4: The principle behind astrometry, a technique for finding extra-solar planets.

Examining two intriguing discoveries

In early 2007, researchers from Switzerland used radial velocity measurements to find the first potentially Earth-like habitable planets beyond the Earth's solar system. These two planets both orbit the red dwarf star Gliese 581, which is just 20 light years from the Earth in the constellation Libra.

The first of the planets, Gliese 581c, is the smallest *exoplanet* (a planet outside our own solar system) ever found within the habitable zone of its star. At first glance, scientists thought the planet looked promising as a possible home for life. But it turns out that the planet is probably too hot for water to be present. Gliese 581c is 14 times closer to its star than the Earth is from the Sun. Even though Gliese 581c's star is a lot cooler than the Sun, the planet is still outside the habitable zone where liquid water can exist.

The second planet is called Gliese 581d. It is larger and farther away from the star than 581c. Gliese 581d seems more promising for the existence of life. Its distance from its star puts Gliese 581d on the outer edges of the habitable zone, but scientists have calculated that the greenhouse effect of gases in its atmosphere may mean that the planet is warm enough for liquid water to flow.

Chasing shadows: The transit method

When planets pass between the Earth and the star they orbit, they offer another faint clue as to their existence – by dimming the amount of light that you can see from the star.

Even the largest planets reduce the brightness of the star by only a tiny amount, but the technology now available to astronomers means that detecting smaller, rocky planets is possible. That's the aim of NASA's Kepler mission, which is planned for launch in 2008 and is going to use a telescope to detect the transits of planets the size of the Earth or smaller around stars in the Earth's region of the Milky Way.

The transit of planets across distant stars tells scientists quite a lot about the nature of that planet. For example, the length of time the planet takes to orbit its sun tells them its mass (based on Kepler's third law of planetary motion; refer to Chapter 3). Scientists can also calculate the planet's size based on how much the planet dims the light of its star.

Based on calculations of the size of the planet's orbit and the temperature of the Sun, researchers can also figure out the planet's temperature. From this figure, they can tentatively answer the question of whether the planet is suitable for habitation similar to the kind of life that exists on the Earth.

The NASA scientists working on the Kepler mission are focusing their attention on the so-called 'habitable zone'. This area is the Goldilocks region around a star where the temperature is not too cold and not too hot, but just right for planets to have liquid water on them – an assumed necessity for life.

Finding Intelligent Life Elsewhere

Even when scientists find potentially habitable planets orbiting other stars using the methods we describe in the preceding section, these discoveries don't guarantee that the planets have life on them – let alone any intelligent life. So if any other civilisations exist, how can humans ever know?

Relying on radio waves

Since the 1960s, one of the most popular techniques in the search for extraterrestrial intelligence (SETI) is to monitor radio waves reaching the Earth from outer space.

Radio waves are a great way to listen for ET phoning home. Like all forms of electromagnetic radiation, radio waves travel at the speed of light and can easily move through the clouds of gas and dust that fill space. Although many things in the universe emit radio waves – such as quasars, pulsars, and even cold hydrogen gas – their natural static tends to be spread out in frequency. People searching for extraterrestrials limit their search to narrow-band signals, which can only be generated by a transmitter.

Using radio telescopes, SETI researchers have been scanning the Milky Way for radio signs since 1960 when the astronomer Frank Drake tried to eavesdrop on interstellar communications with a radio telescope measuring 26 metres in diameter in West Virginia. Drake pointed his telescope at two sun-like stars, but heard nothing that was suggestive of alien life. Still, his work generated a lot of enthusiasm among scientists, and the following year the first major SETI conference was held.

In fact, Drake came up with a formula that combines into one equation many of the unknown factors that are likely to affect the likelihood of finding alien life. Appropriately enough, the formula is known as the Drake Equation, and it looks like this:

$$N = R^{*} \times f_p \times n_e \times f_l \times f_i \times f_c \times L$$

Impressive, huh? What all this means is that the number of civilisations in the Milky Way who are broadcasting in radio waves (N) is equal to:

- The rate at which appropriate, long-lived stars form in the galaxy (R^*)
- Multiplied by the fraction of them that have habitable planets (f_p)
- Multiplied by the number of planets that can incubate life (n_e)
- Multiplied by the fraction that actually develop life (f_l)
- Multiplied by the proportion of those that result in intelligent life (f_i)
- Multiplied by the fraction of intelligent societies that invent technology (f_c)
- Multiplied by the lifetime of societies that use technology (L)

Phew. So what's the answer? How many of those civilisations exist in the Milky Way? Well, no one really knows, because many of the terms in the equation are unknowns. Guesses range from one (the Earth) to a few million.

Several SETI projects have followed Frank Drake's groundbreaking effort. Project Phoenix, for example, was the most sensitive SETI project so far, run by the SETI Institute in Mountain View, California, from 1995 to 2004. Using a handful of telescopes, including the 305-metre diameter Arecibo radio telescope in Puerto Rico, Project Phoenix targeted its search on 750 Sun-like star systems. Unfortunately, no persistent, clearly extraterrestrial signal was found.

In fact, so far, all efforts at tracking down alien civilisations using radio telescopes have failed to turn up anything convincing. That's not to say there haven't been some interesting signals, though. Perhaps the most famous was the so-called 'Wow' signal that was picked up at the Ohio State Radio Observatory in 1977. In August of that year, a volunteer named Jerry Ehman noticed a startlingly strong signal received by the telescope, which he circled on a printout and highlighted with the word 'Wow!' scribbled in the margin. Sadly, despite repeated searches, the signal was never detected again.

Still, SETI advocates are far from giving up the search. The most ambitious project to date, called the Allen Telescope Array, is being constructed by the SETI Institute and the University of California, Berkeley. This array is going to use 350 small telescopes spread over half a mile of California countryside and be dedicated to 24-hour a day, 7 day-a-week searching for alien intelligence (see Figure 17-5).

Journeying the cosmos with Carl Sagan

Perhaps the best known and best loved advocate of the search for extraterrestrial intelligence during the 20th century was Carl Sagan, an astronomer from New York who was a master of extolling the mystery and grandeur of the universe.

Sagan was a consultant to NASA for decades and involved in many expeditions, including briefing the Apollo astronauts before their flights to the Moon. In the wider world, however, he's known for his many books, lectures, and TV shows. In 1980, for example, he wrote and presented an award-winning television series called *Cosmos: A Personal Voyage*, which has been seen by more than 500 million people in more than 60 countries.

Sagan was also closely involved in designing plaques that were sent out into the cosmos on NASA spacecraft, bearing the images of a man and woman and a map of the Earth's location in the galaxy. The plaques are supposed to serve as messages to alien civilisations who may read them one day in the distant future.

As Sagan once said: 'The significance of a finding that there are other beings who share this universe with us would be absolutely phenomenal; it would be an epochal event in human history.'

Figure 17-5: The Allen Telescope Array will consist of 350 telescopes all searching for alien intelligence.

Courtesy of the SETI Institute

Wondering where all the aliens are

One day soon after the end of World War II – probably during the summer of 1950 – the physicist Enrico Fermi paid a visit to some colleagues at the Los Alamos National Laboratory in New Mexico, USA.

As the four of them ate lunch, conversation ranged far and wide, from the possibility of travel faster than light to the disappearance of dustbins from New York City. At one point, apropos of nothing, Fermi came out with a question that has challenged scientists ever since: 'But where is everybody?'

Fermi's lunch companions knew precisely what he was talking about. If the Milky Way galaxy is endowed with so many stars (tens or hundreds of billions of them) and is so old (billions of years), why hasn't a civilisation developed out there somewhere with the capacity to contact the Earth? That's not even mentioning the squillions of galaxies in the wider universe.

The Fermi Paradox, as it's called, highlights the disparity between the fact that aliens have had ample time to colonise the galaxy and the simple truth that Earthlings have absolutely no indication that they're out there.

What does it mean? Are all the alien civilisations out there hiding? Is life on the Earth the only life in the universe? Do technological life forms all end quickly in a moment of self-destruction akin to humans blowing up the world with nuclear weapons? These answers and many more have been offered in response to Fermi's question. Unless someone or something contacts us, we may never be any the wiser about the real answer.

Chapter 18

Coming to an End

In This Chapter

- Predicting the ultimate fate of the Earth and the Sun
- Determining whether the end is going to be cold and dark – or hot and bright
- Ripping apart everything in the universe

Looking into the universe is rather like looking at the announcements page of a local newspaper – you see a number of births, a smattering of deaths, and a handful of marriages. When you look at the night sky, you see stars, galaxies, and gas and dust clouds being born, dying, and colliding with each other to make new entities.

Everything we discuss in this book points to the fact that the universe is changing. In Chapter 7, we explore how the universe itself is expanding. But a lingering question is whether this expansion is going to go on forever or whether it's eventually going to come to a halt under the pull of gravity.

More worryingly, the universe's energy may be drying up. Just as your set of wine glasses eventually ends up broken and ultimately destroyed, the universe is probably condemned to a future where every structure – planets, stars, galaxies – falls apart and becomes colder and darker.

Other theories suggest that the universe may go through an infinite series of cycles of Big Bangs followed by Big Crunches – and that the process that the universe has gone through in the past 14 billion years is eventually going to get played out in reverse.

The fate of the universe may be as far from its origins as you can get, but understanding more about the universe's future may tell a lot about those early days. For this reason, this chapter looks at how everything is likely to end.

All this need not be a reason to get downhearted. The universe might be destined for a fiery or icy end, but either way it probably won't happen for billions of years. We Earthlings have plenty of time to enjoy ourselves, assuming we take good enough care of our little home planet, that is!

Watching the Sun Burn Out

If we don't blow ourselves up or frazzle the planet because of our inability to cut down on carbon dioxide emissions, scientists contend that we are going to be around for some considerable time.

Even if changes to the Sun or bombardment by mega-asteroids do consume the Earth, we can hope that by then science and technology will enable humans to observe the demise of the universe from a vantage point of safety. (We're always looking on the bright side of things, eh?)

Still, one of the first milestones on the timeline of the universe's future is how long the Sun can survive. The Sun is a fairly typical yellowish star and sits on the main sequence of the Hertzsprung-Russell diagram (refer to Chapter 5 for more on this chart that compares a star's luminosity to its surface temperature). Figure 18-1 shows how the evolution of the Sun may go in the years to come.

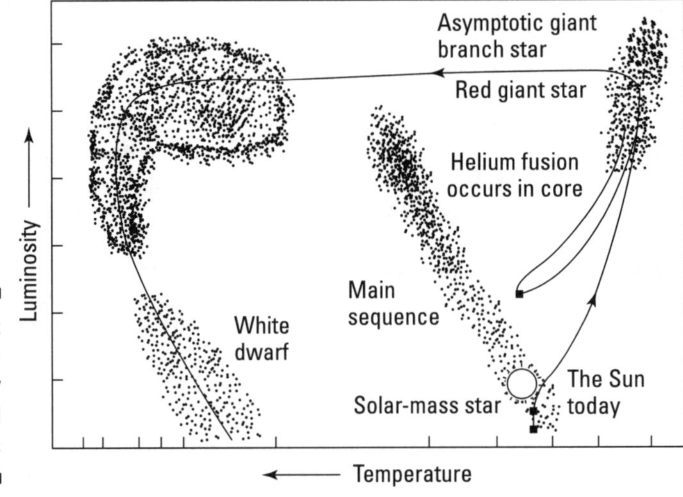

Figure 18-1: How the Sun may evolve until its death.

Before you get worked up, consider that the lifespan of the Sun is estimated at around 10 billion years. Scientists agree that the Sun is around 4.5 billion years old now, so that means it still has plenty of hydrogen left in its core to burn through nuclear fusion for another 5.5 billion years.

But what happens when the Sun's source of fuel does run out? Billions of years from now, after all hydrogen in the Sun's core has been converted to helium through fusion, the pressure won't be high enough to start the fusion of the helium into heavier elements (see Chapter 16).

At this point, hydrogen outside the Sun's core will continue to fuse into helium in spherical shells around the core, and the outer layers of the star will expand outwards massively, turning the star into what is known as a *red giant*. The Sun will now be so huge it will engulf the planets Mercury and Venus and perhaps even the Earth. Even if it doesn't expand this far out, life on the Earth will probably be impossible, as the seas and the atmosphere will boil off into space. This red giant stage of the Sun's new life is estimated to last more than a billion years.

Eventually, the temperature in the core of the red giant will get high enough to allow helium to undergo nuclear fusion into carbon and oxygen. The burning of elements in shells outside the core will continue in an increasingly unstable manner and eventually the outer layers of the star will billow out into space, forming a cloud of glowing gas called a *planetary nebula*. The first planetary nebula was seen by galaxy cataloguer extraordinaire Charles Messier in 1764 – the Dumbbell Nebula.

At this point, the Sun's temperature will never get high enough to start nuclear fusion of the carbon and oxygen, and the core will eventually become relatively inert. The star will then become a *white dwarf*. Even though no nuclear fusion is happening, the star won't collapse completely because of *degeneracy pressure*, which is not like the pressure in a gas, such as the air in your tyre, but a resistance to collapse caused by quantum mechanics.

In Chapter 16, we explain how particles such as electrons can't exist in the same quantum state as other electrons. The collapse of the star tries to force electrons into the same state, but the Pauli exclusion principle keeps the electrons out of each other's space.

The Sun will remain as a cold white dwarf for the rest of its days as a star and continue to cool off – or something more dramatic may happen to the universe as we discuss in the following section.

Contemplating the Fate of the Universe

In Chapter 7, we talk about how the universe is one of three types – open, closed, or flat – depending on the density of matter in the universe. The key difference between the types is what they predict about the ultimate fate of the universe.

- **A closed universe** contains enough matter so that its overall attraction is eventually enough to slow down expansion to a stop and then start to reverse the process, pulling the universe back in on itself.
- **An open universe** doesn't have enough matter, and the universe keeps on expanding in all directions for an infinite amount of time.
- **A flat universe,** which sits on the fence between the extremes of a closed or open universe, allows expansion to eventually come to a stop – but only after an infinitely long amount of time.

Figure 18-2 illustrates what we are talking about in terms of universe types, graphically depicting the average distance between galaxies.

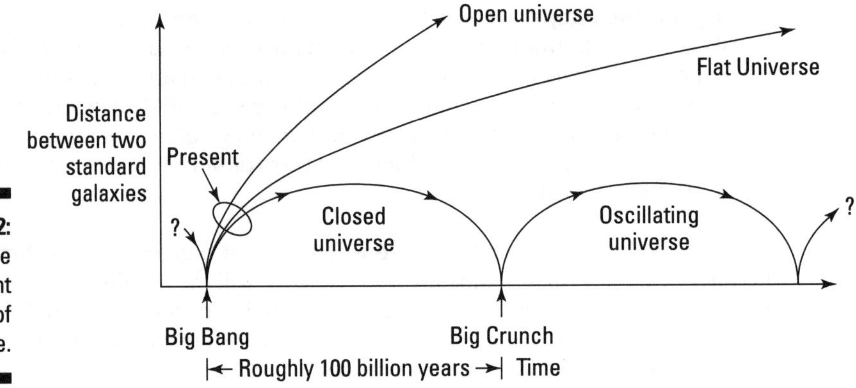

Figure 18-2: The different types of universe.

We discuss the impact of these types of universe in the following sections.

Paying attention to the density of the universe

Much of the discussion about what type of universe humans live in – and by extension its fate – centres on a ratio that scientists refer to using the Greek letter omega, which we discuss in Chapter 7. Omega (Ω) is the ratio of the actual mass density of the universe to the critical density, the boundary between a forever expanding universe and one that contracts under gravity.

Explaining entropy

Omelettes, broken glasses, and hot cross buns seem to have very little to do with the end of the universe, but they may be very relevant indeed.

Suppose that you're making an omelette. You crack a few eggs, whisk them up with some ingredients, and then fry the mixture in some oil in a pan. A few minutes later you have your omelette. If someone asks you to turn the omelette back into its constituent eggs, you may feel compelled to empty the pan's contents over his or her head. Everyone knows that reversing the cooking process is impossible – but why?

Similarly, consider a wine glass that gets knocked off the table and breaks into a thousand pieces. Those pieces can't spontaneously re-form themselves into the unbroken glass. But why?

Finally, think about a hot cross bun, just retrieved from under the grill and placed on a cold plate. After a time, the bun cools down and the plate warms up to some intermediate temperature. Of course, the bun doesn't get heated up again by taking energy from the plate. But why not?

The reason none of these things happen is because of *entropy*, which is a measure of the orderliness of a system of things. The universe is thought to be bound by the so-called *second law of thermodynamics*, which says that entropy can't decrease and is likely to increase as time goes on. There are other laws of thermodynamics – even a zeroth law! – but we don't need to go into those here.

In practice, this means that things like broken glasses, left to their own devices, will not get more ordered. So differences in temperature and lots of other things, such as pressure and *kinetic energy* (the energy due to something's movement), tend to become evened out over time.

How can we know what the critical density is without knowing the mass of everything in the universe? Well, scientists have calculated it by equating a galaxy's kinetic energy (the energy it has due to its velocity through space) and the amount of energy this galaxy needs to escape the gravitational pull of the rest of the mass in the universe.

This sounds like scientists need to know some masses, but it turns out that they don't, because the masses cancel from the equations. In fact, a simple back-of-the-envelope calculation gives an approximate value for the critical density (Ω) as:

$$\rho = \frac{3H_0^2}{8\pi G}$$

In this equation, H_0 is Hubble's constant (refer to Chapter 5), and G is the universal gravitational constant (refer to Chapter 3). Interestingly, everything on the right-hand side of the equation is a constant – which suggests that the critical density value is the same everywhere. However, the answer that the equation gives is only approximate because it doesn't take into account Einstein's general theory of relativity.

Still, the equation does give a reasonable approximation to the expected value for critical density: approximately 10^{-27} kilograms per cubic metre. This number is very small indeed considering that the air we breathe has a density of around 1.2 kilograms per cubic metre. In fact, this critical density equates to just a few hydrogen atoms per cubic metre.

Given that some regions of space have densities much higher than the critical density – in galaxies and black holes, for example – if the universe is to expand forever, some regions of space must be very empty to bring down the average.

The end in an open universe: The Big Chill

If the actual density of the universe is less than the critical density, the universe doesn't contain enough mass to rein in its expansion. So what's going to be the fate of the universe in this case?

If you take a look back at Figure 18-2, you see that it shows the distance between galaxies. In an open universe, galaxies tend to get farther and farther apart as time goes on (apart from any local gravitational influences, for example the gravity experienced between the Milky Way and the Andromeda galaxies).

Based on Hubble's law (refer to Chapter 5), as the distance to a galaxy increases, its *radial velocity* (the speed at which it is moving away from us) also increases. At a point known as *the Hubble Limit*, the radial velocity reaches the speed of light, and humans can no longer observe these galaxies from the Earth. Furthermore, as the expansion continues, at some point the Milky Way (including any galaxies it has merged with by that point) becomes the only galaxy that remains visible from the Earth.

In the sidebar 'Explaining entropy' we talk about, well, entropy. This is a topic covered by the second law of thermodynamics which states that entropy (the tendency for orderliness) is likely to increase over time, which is to say the differences in the universe will tend to get evened out.

In the very, very long run, entropy spells a cold and dark future for the universe, one that cosmologists like to call the *Big Chill* or the *Big Freeze*. Over perhaps trillions and trillions of years in the Big Chill scenario, every

Proton decay

If the prospect of a Big Chill were not enough, another problem may face the universe. Some particle physicists involved in postulating the grand unified theories that we discuss in Chapter 10 believe that the proton, the positively charged particle at the heart of every atom, is inherently unstable.

Some scientists believe that after a certain amount of time all protons will decay into other fundamental particles, probably a pion and a positron (refer to Chapter 9).If this is the case, at some point in the future, every atom is going to fall apart as its constituent protons decay.

So when will this happen? Luckily, even the pessimists say proton decay isn't going to happen for some time – certainly not before 10^{32} years and maybe not as early as 10^{35} years.

Proton decay may not seem like something humans really need to worry about, but it has an important impact on the origin of the universe. Believing that the proton will decay at some point in the far future helps to partly explain why the universe currently contains so much more matter than anti-matter.

Everything scientists know about the interchangeable nature of mass and energy from Einstein and the laws of conservation of certain physical quantities (refer to Chapter 4) naturally leads scientists to also assume that matter and anti-matter were created in equal quantities in the early days. A long but finite lifetime for the proton helps explain why more matter appears to have been created.

star dies. Some go the way of the Sun as a cool white dwarf with no internal energy (see the earlier section 'Watching the Sun Burn Out'), some become cold, dark neutron stars, and other larger stars become black holes (check out Chapter 16 for more on both neutron stars and black holes).

Normally, the remnants of supernovae that create black holes and neutron stars go on to form the breeding grounds for new stars. Eventually, however, theorists believe that the second law of thermodynamics is going to win out and enough free energy to form new stars will not exist. The universe's black holes will swallow up everything that comes into their paths, and the universe will become a very unappealing place to be.

The end in a closed universe: The Big Crunch

If humans live in a closed universe – one whose density is greater than the critical density – a different fate awaits.

If enough matter exists in the universe, the Hubble expansion eventually slows down to a stop, and the galaxies start to fall inwards again. If this is the case, we will be able to observe changes in galactic spectra – where spectra were once red-shifted, they will become blue-shifted.

As a closed universe gets older, more and more galaxies will appear in the observable universe, and these galaxies will get closer and closer to each other. Collisions between galaxies, such as the one predicted to happen between the Milky Way and the Andromeda galaxy at some point, will become increasingly common.

In a closed universe, the sky will also get much brighter. The proximity of all these galaxies means that the sky will blaze with light. The close proximity of other galaxies will also increase our chances of being in the neighbourhood of some pretty nasty supernova explosions.

Eventually, as collisions become more and more frequent, the temperature of the universe will increase to the point where atoms fly apart into their constituent fundamental particles. The processes of the Big Bang may begin to work in reverse until all matter is contained in a singularity once again and the universe goes down perhaps 50 billion years from now in a correspondingly fiery Big Crunch.

What happens next is anyone's guess. Perhaps another Big Bang? Whatever happens, humans won't be around to see it – unless we can escape by using some sci-fiesque spacetime vehicle that takes us into some alternate universe, of course!

Few scientists believe that the Big Crunch is really going to happen to the universe. Observations from satellites like COBE and WMAP show that the universe is most likely open or flat, and not closed.

Considering an Alternative Ending: The Big Rip

In addition to the possibility of a cold, dark ending for the universe or one in which the universe collapses in on itself again, another worrying scenario exists. Scientists aren't always such downers – we promise!

A few years ago, a group of scientists suggested that the universe may be ripped apart by some phenomenon that works to push things away from each other, in other words, *anti-gravity*. This new theory – part of a 2003 scientific paper entitled 'Phantom Energy . . . Causes a Cosmic Doomsday' by Robert Caldwell at Dartmouth College in New Hampshire and Marc Kamionkowski and Nevin N. Weinberg at Caltech in Pasadena, California – postulates the existence of *phantom energy*, a dark energy whose density increases as time goes on.

If phantom energy exists, the end of the universe is going to be rather exciting to witness. Anti-gravitational force will eventually overwhelm every other force in the universe, ultimately ripping everything apart.

In one scenario, Caldwell and his colleagues say that this 'Big Rip' will happen approximately 20 billion years from now. Humans are going to get ample warning that the event is to happen. A billion years before the Big Rip, all the universe's galactic clusters will be ripped apart, although the galaxies themselves will remain – for some time at least.

About 60 million years before the Big Rip, galaxies themselves will be pulled apart. For example, the Milky Way will get ripped apart into its constituent stars. Humans will also be able to see distant galaxies getting torn apart, although the Milky Way's destruction will be seen first because of the delay in receiving light from these distant galaxies.

Our biggest nightmare will come true a few months before the end of the universe when the Earth's solar system gets it, and the planets are pulled out of their orbits. If humanity has survived this long, which seems unlikely, this action almost certainly spells the end for humans when the Sun's energy is taken away.

But even if Earthbound humans do survive the ripping apart of their solar system, they have only a few extra months to go before everything is broken up. The Earth itself will be pulled apart 30 minutes before the end of the universe. A mere fraction of a second (10^{-19}) before the Big Rip, atoms will get pulled apart into their constituent fundamental particles. The cheery scientists end their paper on the Big Rip by saying

> *The current data indicate that our universe is poised somewhere near the razor-thin separation between phantom energy, cosmological constant, and quintessence* (see Chapter 11 for more on these). *Future work, and the longer observations by WMAP, will help to determine the nature of the dark energy. In the meantime we are intrigued to learn of this possible new cosmic fate that differs so remarkably from the re-collapse or endless cooling considered before.*

Of course, the Big Rip is just a hypothesis, and as the quote above makes clear, much more research is needed before scientists are convinced of the definite fate of the universe.

Part V
The Part of Tens

'They still argue about the origins of the universe.'

In this part . . .

This is the part of the book where we give you some snappy material to reel out in those late-night conversations about the meaning of life, the universe, and everything. Here we reveal ten different beliefs and ideas about the origins of the universe, ranging from Terry Pratchett to prayer books.

Also in this part we explain the big-ticket experiments that cost gazillions of dollars to implement but that have helped humanity advance on the road towards cosmic understanding. If you don't know your Kecks from your COBE, this is the place to look.

Confused by Fahrenheit, Celsius, and Centigrade? Scientists like to confuse things even more with their own temperature unit – the kelvin. The Appendix is where you can find out what it's all about. We introduce you to a lot of other technical material, but without any of that baffling boffinese.

Chapter 19

Ten Different Beliefs about the Origins of the Universe

In This Chapter

- Explaining the universe from a religious perspective
- Telling stories about the universe in folklore
- Laughing over fictional accounts of the universe's beginning

As we say a few times in this book, an overwhelming majority of scientists believe that the universe started something like 14 billion years ago with a hot Big Bang and has been expanding ever since. This explanation fits with all the evidence they've gathered – from observing the stars to probing the secrets of the atom.

Of course, science isn't alone in offering answers to the big questions of creation. Religions have their own answers, rooted in systems of belief that are often millennia old. Even more ancient in some cases are the creation tales told in the folklore of tribal societies. We offer a glimpse into some of these beliefs in this chapter.

We also talk in this chapter about three accounts of the start of the universe that come from fiction. The topic of how the universe came to be is almost too much for fiction writers to ignore.

By commingling traditional beliefs of various peoples, societies, and cultures with works of fiction in this chapter, we're not implying that the religious beliefs and tribal folklore are in any way imaginary or untrue. This chapter is simply a listing of ten ways of looking at the origins of the universe that differ from cosmologists' views.

Judeo-Christian Creation: In the Beginning

In the Jewish and Christian faiths, the story of the creation of the universe is told in the book of Genesis, in the Torah or Old Testament of the Bible. In the first 19 verses of the first chapter of the book, God creates the universe step by step.

First, light is brought into being and separated from the darkness. Then Heaven is divided from the Earth and seas. On the third day, Genesis tells of God focusing his attention on the plants of the Earth before, on the fourth day, he creates stars, the Moon, and the Sun to give light to the Earth.

To get from formless void to a functioning creation in four days certainly tells a different story from that offered by cosmologists. Belief in the literal truth of the creation story in Genesis leads some people to calculate the age of the Earth in thousands of years, rather than billions. In the 16th century, for example, James Ussher, Archbishop of Armagh in Ireland, used the chronologies and genealogies of the Bible to calculate the date of creation at 23 October 4004 BC. More recent calculations along similar lines put the age of the Earth at somewhere closer to 6,000 years.

Islamic Creation: Opening the Heavens and Earth

In the Koran, the holy book of the Islamic faith, the creation of the universe is mentioned in several places. In the 41st chapter of the Koran, for example, Allah's formation of the world is described as follows:

> *He directed Himself to the heaven and it is a vapour, so He said to it and to the earth: Come both, willingly or unwillingly. They both said: We come willingly.*
>
> *So He ordained them seven heavens in two periods* [days], *and revealed in every heaven its affair; and We adorned the lower heaven with brilliant stars and [made it] to guard; that is the decree of the Mighty, the Knowing.*

As with the biblical view of creation, everything is made in a period of six days: two days to make the Earth, two days to make the mountains, and two days to create seven heavens, through which Muhammad was later to journey on his way to paradise. The lowest of these heavens contains the planets.

Another overview of creation in the Koran describes the origins of the universe like this:

> *Do not those who disbelieve see that the heavens and the earth were closed up, but We have opened them; and We have made of water everything living, will they not then believe? . . . And We have made the heaven a guarded canopy . . . And He it is Who created the night and the day and the sun and the moon; all [orbs] travel along swiftly in their celestial spheres.*

Hindu Creation: Cycles upon Cycles

Numerous creation stories appear in Hindu sacred texts, but the big picture is this: Hindus believe that the universe goes through cycles of creation and destruction.

The most commonly held belief is that the universe and all human and animal life was created by the Hindu deity, Brahma, who himself was born out of a lotus flower or a golden egg, depending on which account you read. A second deity, Vishnu, maintains the universe on a day-to-day basis, while a third, Shiva, destroys the universe in a cataclysm of fire and water every 4 billion years. Then the whole process begins again.

In the *Rigveda*, a collection of hymns dedicated to the gods that is one of four sacred Hindu books, creation is described in these mystical terms:

> *Then was not non-existence nor existence: there was no realm of air, no sky beyond it. What covered in, and where? and what gave shelter? Was water there, unfathomed depth of water? Death was not then, nor was there aught immortal: no sign was there, the day's and night's divider. That One Thing, breathless, breathed by its own nature: apart from it was nothing whatsoever. Darkness there was at first concealed in darkness this. All was indiscriminate chaos. All that existed then was void and formless: by the great power of Warmth was born that Unit. Thereafter rose Desire in the beginning, Desire, the primal seed and germ of Spirit . . . There were begetters, there were mighty forces, free action here and energy up yonder. Who verily knows and who can here declare it, whence it was born and whence comes this creation?*

Good questions, indeed!

Buddhist Creation: Cause and Effect without a Creator

In common with Hindus, Buddhists think of the universe going through a continuous cycle of creation. In the Buddhist scripture *Agganna Sutta*, a monk Vasettha is told by the Buddha how this happens.

> *There comes a time, Vasettha, when, sooner or later after a long period, this world contracts. At a time of contraction, beings are mostly born in the Abhassara Brahma [the radiant] world. And there they dwell, mind-made, feeding on delight, self-luminous, moving through the air, glorious – and they stay like that for a very long time. But sooner or later, after a very long period, this world begins to expand again. At a time of expansion, the beings from the Abhassara Brahma world, having passed away from there, are mostly reborn in this world.*

When this happened to the world it became a vast dark body of water, with no day or night. The scripture continues:

> *Sooner or later, after a very long period of time, savoury earth spread itself over the waters where those beings were. It looked just like the skin that forms itself over hot milk as it cools. It was endowed with colour, smell, and taste. It was the colour of fine ghee or butter and it was very sweet, like pure wild honey. Then some being of a greedy nature said: 'I say, what can this be?' and tasted the savoury earth on its finger. In so doing, it became taken with the flavour, and craving arose in it. Then other beings, taking their cue from that one, also tasted the stuff with their fingers. They too were taken with the flavour, and craving arose in them. So they set to with their hands, breaking off pieces of the stuff in order to eat it. And the result was that their self-luminance disappeared. And as a result of the disappearance of their self-luminance the moon and the sun appeared, night and day were distinguished, months and fortnights appeared, and the year and its seasons. To that extent the world re-evolved.*

Some take this passage to be an accurate description of the oscillating universe theory – a Big Crunch followed by a Big Bang ad infinitum – although many remain skeptical.

The Tibetan spiritual leader Dalai Lama addresses the question of the origin of the universe in his book on science and Buddhism called *The Universe in a Single Atom*. He says: 'Buddhism and science share a fundamental reluctance to postulate a transcendent being as the origin of all things.' The Dalai Lama explains that during the void 'the particles of space subsist, and from these particles the new universe will be formed'.

The Buddhist view of the universe doesn't include a divine creator. In common with the scientific view, Buddhism accounts for the cosmos 'in terms of the complex interrelations of the natural laws of cause and effect', the Dalai Lama says.

Shinto Creation: The Earth, Young and Oily

Japan's native religion, Shinto, calls on its adherents to live 'a simple and harmonious life with nature and people'. One of its most ancient texts, called the *Kojiki*, describes the beginning of things – albeit in a manner that focuses more on the names of the deities involved than the nitty-gritty of cosmogenesis!

According to the *Kojiki*:

> *The names of the Deities that were born in the Plain of High Heaven when the Heaven and Earth began were the Deity Master-of-the-August-Centre-of-Heaven, next the High-August-Producing-Wondrous Deity, next the Divine-Producing-Wondrous Deity. These three Deities were all Deities born alone, and hid their persons.*

The *Kojiki* goes on to name the next group of deities, who were born 'from a thing that sprouted up like unto a reed-shoot when the Earth, young and like unto floating oil, drifted about medusa-like'. For those who are interested, their names are Pleasant-Reed-Shoot-Prince-Elder Deity and Heavenly-Eternally-Standing Deity.

African Folklore: Egg-centric Origins

The people of southern Mali who speak the Mande languages tell a folktale of how the universe emerged from a kind of cosmic egg (not so different from Georges Lemaître's concept that led to the idea of the Big Bang – refer to Chapter 6). The story begins with a powerful being called Mangala, who had a round, energetic form that was divided into four parts.

Those parts symbolised the four elements (matter), and the four points of the compass (space). But at some point, Mangala had enough of keeping all the universe's matter inside, and so he made it into a seed.

But the seed blew up, and Mangala started again, this time with eight sets of twin seeds, which he planted in an egg-shaped womb where they gestated and turned into a fish, a symbol of fertility.

Eventually, the contents of the universe's womb rebelled, with one male character called Pemba throwing out part of the womb's placenta to form the Earth. Through a series of dramatic episodes involving incest, drums, hammers, and skulls, the tale eventually describes the arrival of night, fire, and the establishment of society – all emanating from this kind of cosmic egg.

Iroquois Creation: The Turtle Time Story

Native Americans have plenty of creation stories to offer. One particularly nice one comes from the Iroquois tribe in the eastern United States and involves clouds, the sea, a tree – and a turtle.

A few different versions of the story exist, but they all share the same basic elements. They begin in faraway days, when the Sky people lived a blissful life on a floating island upon which grew one grand tree. One day, the people tore the great tree out of the ground. In the resulting hole, they were able to see a great cloud sea.

A pregnant woman was sent into the hole, which caused consternation among the animals and birds living in the cloud sea. Two birds caught the woman, and then a series of animals tried to find a way to help her by swimming to the bottom of the sea to bring up mud to make a place where she could live.

One after the other, the animals tried and failed to reach the mud until finally the toad, or perhaps the muskrat (we've seen versions involving both), succeeded.

The question then arose as to who was going to carry the Earth upon which the woman would live. The turtle was willing, and so the Earth was placed on the turtle's shell and quickly grew to the size of the North American continent.

The woman then created the stars, the Moon, and the Sun before giving birth to twins – Flint, a hard-hearted god who put the bones in fish and the thorns on berry bushes, and Sapling, the god who created everything useful to humans.

Adams: Life, the Universe, and Everything

The science of cosmology, religious texts, tales from folklore – and even science-fiction novels – all attempt to answer some of the deepest questions imaginable with varying degrees of levity – how did humans, and everything around us, come into being? Why is everything the way it is? What's it all about?

Well, the very funny *Hitchhiker's Guide to the Galaxy* novels by Englishman Douglas Adams offer a very simple answer. What is it? Well, here's the story first.

According to the books, a civilisation of extremely advanced and intelligent aliens set out one day to answer the ultimate question of Life, the Universe, and Everything, by building an enormously powerful computer called Deep Thought. For millions of years, the computer pondered the question before finally and with much fanfare revealing the answer: 42.

Unsurprisingly, the alien civilisation was less than thrilled with this answer, but the computer stood its ground. The number 42 was definitely the answer, it said – the problem was, the aliens didn't know the question.

So the alien civilisation went back to the drawing board, creating an even more powerful computer in the form of the Earth and all the creatures on it. This computer (the Earth) is about to spit out the question when it is destroyed by another alien life form building a hyperspace by-pass.

Luckily, Arthur Dent, one of two remaining earthlings, manages to escape the destruction by thumbing a lift with an alien, leading to adventures in several other alternate Earths, giving hope to those who seek the ultimate question.

The final book in what Adams called his 'five book trilogy' ends with the destruction of all these alternate Earths. The death of author Adams in 2001 means that the ultimate question of Life, the Universe, and Everything may now never be revealed.

Pratchett: Absurdity and Another Giant Turtle

Ever since novelist Terry Pratchett began writing his comic fantasy *Discworld* series back in 1983, millions of readers have relished his ability to satirise our society by telling the stories of the magical and odd inhabitants of his imaginary one.

The very opening page of the first *Discworld* novel, *The Colour of Magic*, plays this sort of game with cosmology:

> *In a distant and second-hand set of dimensions, in an astral plane that was never meant to fly, the curling star-mists part . . . See . . . Great A'Tuin the Turtle comes, swimming slowly through the interstellar gulf, hydrogen frost on his ponderous limbs.*

The great turtle turns out to be carrying four giant elephants on its back, 'upon whose broad and star-tanned shoulders the disc of the World rests'.

Reading Pratchett's mocking take on cosmology is lovely. He conceives two possible futures for the universe involving the turtle (A'Tuin):

> *Would A'Tuin keep walking until he crawled at a steady gait until he returned to the nowhere he came from? Or was he heading toward a Time of Mating, where he and all the other turtles carrying stars in the sky would briefly and passionately mate, for the first and only time, creating new turtles and a new pattern of worlds? This was known as the Big Bang hypothesis.*

In the World Before Monkey

For British and Australian children growing up in the 1970s and 1980s, one of the highlights of watching television was a cult programme called *Monkey.* In the tale, a monkey hero, a pig monster, and a river ogre (all accompanying a young Buddhist monk riding a magical horse) set off to India to obtain Buddhist religious texts. Along the way, they battle innumerable monsters and generally have a rollicking time.

The opening of the show, which in fact is based on a novel published in the 1590s in China, sets out a dramatic origin for the world: Out of chaos, a certain stone egg is formed. To get the sense of it all, recite the following voiceover from the beginning of the show in your most exotic accent, building to a crescendo on the last sentence!

> *In the world before monkey, primal chaos reigned. Heaven sought order, but the phoenix can fly only when its feathers are grown. The four worlds formed again and yet again as endless aeons wheeled and passed. Time and the pure essences of heaven all worked upon a certain rock, old as creation. It became magically fertile. The first egg was named 'Thought'. Tathagata Buddha, the Father Buddha, said, 'With our thoughts we make the world.' Elemental forces caused the egg to hatch. From it came a stone monkey. The nature of monkey was irrepressible!*

Fans can still catch the series on DVD or on a large number of fan Web sites.

Chapter 20

Ten Greatest Cosmological Advances

In This Chapter

- Identifying exceptional experiments
- Highlighting amazing institutions
- Profiling prodigious projects

Human understanding of the origins of the universe comes from many sources, but some scientific institutions have hit above their weight in relation to cosmology.

This chapter is our top ten list of telescopes, particle physics laboratories, and satellites that have helped the most in delving farther back in time.

Cosmic Background Explorer (COBE)

Although space missions have revealed much about the origins of the universe, the launch of COBE in 1989 was the first time that cosmologists had their own satellite dedicated to the study of cosmology. COBE's task, until the mission ended in 1993, was to analyse the cosmic background radiation discovered in 1965 by Arno Penzias and Robert Wilson (refer to Chapter 6 for more on cosmic background radiation).

NASA's Goddard Space Flight Center developed the satellite, which included three different instruments with specific functions:

- **The Diffuse Infrared Background Experiment (DIRBE)** looked for cosmic background radiation in the infrared part of the spectrum. The instrument effectively took a sample dating back to the early universe. Its measurements enabled cosmologists to better understand how stars are formed and how chemical elements heavier than hydrogen have built up throughout the universe.
- **The Differential Microwave Radiometer (DMR)** mapped any changes in the cosmic background radiation across the sky. Its biggest achievement was to find small variations in the temperature (about a hundred thousandth of a degree) of the radiation. These variations – though scientists still don't know exactly how – developed into the galaxies and clusters you see today.
- **The Far Infrared Absolute Spectrophotometer (FIRAS)** measured the spectrum of the cosmic microwave background radiation and compared it with the expected spectrum from a black body (described in Chapter 9) with a temperature of 2.725 +/–0.002 kelvin, which was the temperature predicted by Big Bang theorists. It found an almost perfect correlation between the two and led cosmologists to believe that nearly all the measured radiation was released within a year of the Big Bang happening (refer to Chapter 6 for more about the Big Bang).

For more on COBE, visit this NASA Web site: `http://lambda.gsfc.nasa.gov/product/cobe`.

European Particle Physics Laboratory (CERN)

The laboratory known as CERN (an abbreviation of the French name for the facility, if you're wondering) has given the world many things. For example, it's the location where, in 1989, Sir Tim Berners-Lee first invented the World Wide Web (although not the Internet generally, which had been around in US military and academic circles for decades before).

Located in Geneva on the border of France and Switzerland, CERN has made perhaps its greatest discoveries in the area of particle physics.

- In 1973, André Lagarrigue and colleagues found strong evidence in support of *electroweak theory* – the concept that electromagnetism and the weak force were different aspects of the same thing (flip to Chapter 10 to read more about electroweak theory). Using CERN's Gargamelle bubble chamber, the team captured images of tracks that showed particle interactions involving so-called *neutral currents*.

- In the following decades, CERN upgraded its *particle accelerator*, which sped up elementary particles and smashed them into fixed targets, into a *particle collider*, which smashes intense beams of protons and anti-protons into each other. The extra energy that this created enabled scientists, under the guidance of CERN Director General Carlo Rubbia, to directly observe the particles that transmitted the weak force – the W and Z bosons.

These discoveries and others are critical in understanding the origins of the universe. Some theories predict that all the fundamental forces of nature – electromagnetism, weak, strong, and gravity – are just four aspects of the same thing and that in the early moments of the universe they were unified. As CERN's experiments reach higher energies, scientists are essentially probing back farther and farther in time.

CERN's next big experiment is the Large Hadron Collider (LHC). The laboratory has rebuilt the 25-kilometre long particle collider, a structure it bills as the world's largest machine. The collider, which sits under the Jura Mountains close to the French-Swiss border, is going to smash two beams of protons together at energies of up to 14 TeV (tera-electronvolts, or trillion electronvolts). When scientists smash protons together at this energy, they should create a massive burst of elementary particles. Among these created particles, CERN scientists hope to find evidence of the Higgs boson and perhaps supersymmetric particles. (See Chapters 10 and 11 for more on these still theoretical particles).

The official CERN Web site, `www.cern.ch`, contains a wealth of information about the LHC and other cool stuff.

Hubble Space Telescope

Of all the experiments we list in this chapter, few have a greater public profile than the Hubble Space Telescope, the world's first space-based optical telescope. If you read newspapers or magazines or spend much time on the Web, you've probably seen at least one of the spectacular images produced by this super-telescope.

Hubble is an absolute monster. The spacecraft measures the length of a school bus and its main mirror is 2 metres 31 centimetres in diameter. Clocking in at 11.1 tonnes, the spacecraft weighs as much as two elephants.

Elephants are a sore point at Hubble command headquarters, however, because the telescope may easily have become a white elephant after its main mirror was found to have a serious design flaw immediately after its launch in

1990. Fortunately, space shuttle astronauts in 1993 used a number of clever techniques to restore Hubble's sight.

Hubble's power comes from its ability to observe the skies in visible and ultraviolet light with incredible clarity because it doesn't have to view objects through the turbulent atmosphere of the Earth.

Hubble has an incredible catalogue of discoveries to its name:

- Observing planets orbiting distant stars
- Showing star nurseries (refer to Chapter 13) in unprecedented detail
- Refining the value of the Hubble constant (check out Chapter 5) to a greater accuracy than any previous experiment
- Leading scientists to realise that the universe's expansion is accelerating, not slowing down

Yet Hubble's day will soon be over. It's set to be replaced by the James Webb Space Telescope as early as 2013. This new instrument is going to have ten times the light-gathering power of Hubble.

To see some of the beautiful Hubble images, visit `http://hubblesite.org/gallery`. Many of Hubble's best images are available to view in the new Google Sky repository at `http://earth.google.com/sky`. You can find out more about the James Webb telescope at `www.jwst.nasa.gov`.

Super-Kamiokande

Super-Kamiokande is a joint neutrino-detecting experiment between the USA and Japan. The project takes its name from the Kamioka Mine in which it's situated, some 250 kilometres north of Tokyo, and the initials for nucleon decay experiment.

Like many other neutrino experiments, the set-up at Super-Kamiokande includes a huge tank of liquid – in this case a tank of ultra-pure water that measures some 40 metres across by 40 metres tall. This much water weighs a massive tens of thousands of tonnes. To avoid contamination from cosmic rays, the tank is a kilometre underground.

Surrounding the water are more than 11,000 *photomultiplier tubes* – detectors that 'amplify' the telltale Cerenkov radiation produced when neutrinos interact with particles in the water. *Cerenkov radiation* is emitted when a particle travels faster than the speed of light for the material in which it is travelling. In water, the speed of light is slower than in air.

Super-Kamiokande's first big results came in 1987 when the detector observed a burst of neutrinos produced by the massive supernova SN1987a in a nearby dwarf galaxy. The following year, scientists observed neutrinos coming from the Sun.

Perhaps the most important result was the observation of so-called *neutrino oscillations*. We explain in Chapter 9 how neutrinos come in three different flavours (electron, mu, and tau). Theorists predicted that if the neutrino had mass – rather than being massless, as originally thought – the neutrino could flip among these different flavours. Super K observed just this flipping between flavours and provides some of the strongest evidence that the neutrino does have mass.

Super-Kamiokande is also being used to look for proton decay (which we discuss in Chapter 18), although no evidence of this has yet been seen.

Wilkinson Microwave Anisotropy Probe (WMAP)

The name of this NASA satellite, launched from Cape Canaveral in 2001, is something of a mouthful, but its original purpose was to map the variations in the cosmic microwave background first spotted by the DMR instrument on the COBE mission. See the earlier section 'Cosmic Background Explorer (COBE)'.

The satellite is named after Princeton's David Wilkinson, who was one of the originators of the WMAP mission and who died in 2002. The satellite sits a million miles from the Earth and orbits around a gravitationally stable position in the Earth-Sun-Moon system known as a *Lagrange point*. This position and a protective shield that points in the opposite direction to the on-board instruments enable the satellite to look out into deep space without interference from the Earth, the Moon, and the Sun.

The big difference between COBE and WMAP is the detail with which it can see variations. COBE could only measure variations in chunks of sky some 14 times larger than the apparent size of the Moon. (Not very sensitive, but it still produced some incredible results.) By contrast, WMAP is more than 20 times more sensitive.

WMAP may also discover a lot about the cosmic background radiation's *polarisation*, the direction in which the electromagnetic wave vibrates (and the scientific concept behind sunglasses that react to sunlight).

Among WMAP's results is convincing evidence for inflation theory (refer to Chapter 7) and the existence of dark matter and dark energy (see Chapter 11 for more on dark matter and energy, and the colour section for an image of how they are distributed throughout the universe). WMAP has also been able to probe to within a trillionth of a second of the Big Bang by comparing the size and brightness of the variations in the cosmic wave background.

The mission has been considered so successful that its initial two-year mission, set to end in 2003, has been extended to 2009. WMAP's results continue to strongly shape theories of the origins of the universe.

Cool pictures and the history of the WMAP project are on display at `http://map.gsfc.nasa.gov`.

Chandra X-ray Observatory

Although the early observations of the sky relied on visible light, astronomers now realise the value of using telescopes that can tune into other sorts of electromagnetic radiation, such as X-rays.

As well as their use in healthcare, X-rays can tell scientists a great deal about some of the oddest things in the universe – black holes, quasars, and supernovae, to name but a few. At the cutting edge of X-ray astronomy is the Chandra X-Ray Observatory, a satellite-based telescope launched in 1999, which provides some of the most stunning images of the universe. Take a look at `http://chandra.harvard.edu/photo`.

Chandra is in a highly elliptical orbit that takes it out to some 139,100 kilometres from the Earth at its farthest point (about a third of the distance to the Moon). The instrument needs to be in space because X-rays would be severely absorbed by the Earth's atmosphere.

Chandra is the largest satellite ever launched by the space shuttle, largely because of the design of its mirrors, which are shaped like long barrels. This design focuses high-energy X-rays and forces them to bounce off the mirrors within the tubes like speeding bullets.

On board Chandra, named after Nobel prize-winning astrophysicist Subrahmanyan Chandrasekhar, are four instruments that precisely measure the number, position, energy, and time of arrival of X-rays that enter the telescope. An on-board high-resolution camera can make out details as fine as being able to read a newspaper from a distance of about a kilometre.

Among the many discoveries made by Chandra, perhaps the most exciting is the observation of X-ray emissions from an object known as Sagittarius A, the supermassive black hole that cosmologists believe lurks at the centre of the Milky Way. The telescope has also been used to study what's going on in star nurseries, leading scientists to better understand stellar evolution (refer to Chapter 13).

Recently, Chandra has been turning its attentions to dark matter (see Chapter 11). In 2006, it found strong support for dark matter's existence by watching galactic clusters merging.

Fermilab

Until the Large Hadron Collider starts (see the earlier section 'European Particle Physics Laboratory (CERN)'), the title of the world's most powerful particle collider goes to the Tevatron at Fermilab in Illinois, USA. Although the collider has smashed protons into antiprotons, producing energies of 1 TeV (a trillion electronvolts) – hence the collider's name – Tevatron is just one in a long line of experiments at Fermilab.

In 1977, Leon Lederman worked on the E288 collaboration to discover a new particle called the *upsilon*, at that time the heaviest sub-atomic particle ever found. New particles had been found before, but this was the first to contain a bottom quark (and an anti-bottom quark for good measure). This discovery validated one of the key planks in the Standard Model of particle physics (Chapter 9 has more on the Standard Model).

Not content with discovering one quark, Fermilab scientists went on to discover

- **The top quark.** In March 1995, scientists on the CDF and DZero experiments on the Tevatron announced that they had observed the telltale signs of the top quark on several occasions. These signs consisted of a muon, a neutrino, and four tightly bunched jets of other particles, which are produced when the top and anti-top quarks produced in the collision very quickly decay into other particles.
- **The tau neutrino.** In July 2000, Fermilab's researchers found yet another piece of the Standard Model jigsaw puzzle. In 1997, the lab had turned on an intense beam of neutrinos and bombarded a detector made up of

photographic emulsion sandwiched between iron plates. It took three years for scientists to examine more than 6 million images of tracks left by the neutrinos to find just four that showed firm evidence of this most exotic of the neutrinos.

To find out more about Fermilab and some of the basics of particle physics, visit `www.fnal.gov`.

Atacama Cosmology Telescope (ACT)

The Atacama Desert in Chile is home to one of the world's newest observatories – and also its highest.

Sitting on top of the Cerro Toco mountain at 5,100 metres above sea level to reduce the effects of water vapour in the atmosphere, the Atacama Cosmology Telescope (ACT) has equally lofty objectives: It wants to map the cosmic microwave background of the universe to even better resolution than COBE or the Planck satellite (due to launch in 2007).

ACT measures some 6 metres across and sits on a turntable so that scientists can rotate the whole instrument to look at different parts of the sky. Unlike COBE, which was a satellite and also looked at the whole sky, the earthbound ACT plans to study smaller sections of the sky in greater detail, through observations lasting months at a time. Although the telescope received so-called 'first light' in June 2007, its main instrument, the Millimeter-Wave Bolometric Camera, won't be operational until later.

ACT's instrumentation has been fine-tuned to view the sky at three specific microwave frequencies in an effort to probe the very earliest moments of the universe and to see how this led to the development of its observable current-day structure. The telescope is going to look for the effects of gravitational lensing of microwaves to do so (see Chapter 4 for more on gravitational lensing.)

The telescope is also going to carry out a study of the very largest galactic clusters and accurately determine the red-shifts of more than 400 of them.

Mount Wilson Observatory

The scientific institution with arguably the greatest impact on cosmology is the Mount Wilson Observatory, founded in 1904 in the San Gabriel Mountains of Southern California. In the first half of the 20th century, the Mount Wilson Observatory was home to the world's largest telescopes – the 152-centimetre Hale telescope and the 250-centimetre Hooker telescope.

Mount Wilson's list of discoveries is hard to beat. Edwin Hubble worked here and demonstrated that distant galaxies are receding from the Earth – the first piece of evidence for the Big Bang theory of the origins of the universe. Hubble also showed from his work at Mount Wilson that the Milky Way is just one of countless galaxies in the universe.

Observations at Mount Wilson also proved that the Sun isn't at the centre of the Milky Way, that the Sun has a magnetic field, and that the stars of the Milky Way have different ages. Not a bad haul, eh?

Work hasn't stopped at Mount Wilson. A new network of six 101-centimetre telescopes dotted around the mountain has just been completed, forming what is known as the CHARA Array. Light from the six telescopes is carried through vacuum tubes to a central location and combined to produce images of incredible resolution. This technique allows astronomers to see extremely fine detail. The array will be able to distinguish individual members of binary star systems (which contain two stars) and more accurately measure star diameters. The Mount Wilson site at `www.mtwilson.edu` is worth a visit.

Keck Telescopes

Although the most powerful Earth-based telescopes of the last century were at Mount Wilson, in the 21st century they are perched on top of a dormant Hawaiian volcano – the 4,200-metre-high Mauna Kea.

Hawaii may sound like an excuse for astronomers to have a nice holiday at someone else's expense, but they have a good reason for siting a telescope here: The ocean surrounding Hawaii is very *thermally stable* (meaning very few atmospheric effects as a result). Also, no other mountain ranges are nearby to throw up dust and very few nearby cities spew out light pollution. Oh, and the weather is just about perfect.

Mauna Kea is home of the WM Keck Observatory and its two 10-metre diameter telescopes Keck I and Keck II. (They had to have some name, we suppose.) The Keck telescopes are best suited for making observations in the visible and infrared parts of the electromagnetic spectrum. Instead of being made from single pieces of glass, the mirrors at the heart of the Keck telescopes are made from 36 interlocking hexagonal mirror segments.

Observations started in 1993, and the scientific results haven't stopped rolling in. Astronomers have seen evidence of planets outside the Earth's solar system, witnessed a collision between two galaxies with supermassive black holes at their centres, and taken some of the best images of the neighbouring Andromeda galaxy. In 2007, scientists used the Keck telescopes to see the oldest observed galaxies, dating back to 500 million years after the Big Bang. Check out `www.keckobservatory.org` for some images from the telescopes.

And the Keck Observatory excels in more than just cosmology. In 2005, observations with the two Kecks helped astronomers to discover a faint ring around Uranus. They even managed to announce this discovery to the world without sniggering.

Appendix

Understanding Scientific Units and Equations

To understand some of the maths in this book, you need to know some basic concepts that mathematicians (and cosmologists) use when writing out numbers and equations.

Powers

Because scientists often encounter very large and very small numbers, they have come up with a nifty form of shorthand for writing them.

In your school maths lessons, you certainly came across the concept of squaring a number. '2 squared' is the same as 2 multiplied by 2. Mathematicians write this as 2^2. You may also have heard about cubing a number. For example, 2 cubed, 2^3, is the same as 2 multiplied by 2 multiplied by 2 again.

In fact, scientists and mathematicians don't stop at cubing a number. You can multiply a number together as many times as you like. Scientists and mathematicians use the term 'to the power *x*', where *x* is the number of times that you want to multiply the number together. So, 2 to the power 10, which can be written as 2^{10}, is 2 multiplied by itself 10 times.

Powers get useful in expressing very large numbers. You know that the number 100 is the same as 10 squared or 10^2. Similarly 1,000 is 10^3, 10,000 is 10^4, 100,000 is 10^5, and 1,000,000 is 10^6. But when some of the things you are studying involve writing out numbers with 20 or 30 zeroes after them, using powers is just plain easier. So rather than writing 1,000,000,000,000,000,000,000, you can write 10^{21}.

Inside the book you see numbers written out like this: 1.3×10^{21}. This layout is *scientific notation* and it works like a little sum. You know what 10^{21} looks like, so just multiply that by 1.3 to get 1,300,000,000,000,000,000,000. The number before the multiplication sign can have as many decimal places as you choose.

A form of this notation using negative numbers can also be used for really small fractions. A half, ½, can also be referred to as 2 to the power of –1 or 2^{-1}. Similarly a quarter can be thought of as 1 divided by 2 squared or 2^{-2}.

The same goes for powers of 10. 10^{-2} means 1 divided by 10 squared. 10^{-3} means 1 divided by 10 cubed. You can therefore write any very small number using these powers of 10. 1.3×10^{-21} is therefore the same as writing 1.3 times 1 divided by 1,000,000,000,000,000,000,000 or 0.0000000000000000000013.

Other Mathematical Conventions

Scientists like to dispense with the multiplication sign in equations and formulae, partly because they want to write things down quickly but also because it avoids confusing the multiplication sign with the letter x, which is often used to specify a distance or unknown quantity.

In Einstein's equation $E = mc^2$, for example, notice that on the right side ol' Albert missed out a multiplication sign between the m and the c.

In mathematical equations in this book and in science generally, the letters used for quantities that can change are written in italic type whereas letters for quantities that are *fixed* (sometimes known as constants) are written in normal upright type. For example, the equation $E = m\mathrm{c}^2$, is written this way because the energy and mass change depending on which particle you're talking about, while c, the speed of light, has a fixed value.

The symbols we use for scientific units (see the following section) are also written with upright letters. For example, the letter m represents the metre and the letter s represents the second in equations.

Scientific Units

The term units refers to how you measure things. Centimetres, metres, inches, feet, kilograms, pounds, ounces, and seconds are all examples of units that you use in everyday life.

Modern-day scientists use a system of units known as *SI*, or *Système International* (meaning 'international system' in French). SI is based on the metric system, which has its roots during the French Revolution. In 1799, two pieces of platinum were deposited at the Archives de la République in Paris. One of them was 1 metre in length, the other weighed 1 kilogram. Prior to that, virtually everyone had a different idea of how long a metre was and how heavy a kilogram was. But from that day in 1799 onwards, people were able to refer to these standard measures of length and weight to ensure that everyone was talking about the same thing.

Since then SI has developed into the universal language of scientists, engineers, and mathematicians. As well as the metre and the kilogram, the system now includes five other units of measure: the second (time), the ampere (electric current), the kelvin (temperature), the mole (amount of substance), and the candela (luminosity).

We don't use all these units in the book but it's useful to know them. In the following sections we look at how the units that appear in this book are defined.

The metre

The platinum bar held in Paris that defined the metre was meant to be one ten-millionth of the distance from the equator to the North Pole as measured through Paris (*mais oui!*). Because scientists now know that the size of the Earth changes due to geological processes and don't want to be too aligned with one country's ideas, scientists have discarded defining the metre from the length of the platinum bar.

The metre is now defined as the length of the path travelled by light in a vacuum during a time interval of 1/299,792,458 of a second. Because the speed of light in a vacuum is fixed, this fixes the definition of a metre.

The kilogram

Unlike the metre, scientists still rely on a block of platinum (mixed with a hint of the metal iridium) for the kilogram. The current block of metal was created in the 1880s and is held in a controlled vault at the International Office of Weights and Measures (or BIPM, abbreviated from the French name), the organisation responsible for maintaining and developing SI.

The second

Defining the second based on the speed of light may seem tempting, but doing so would end up in a serious case of circular logic. Instead, the second is defined from looking at atoms of the chemical element caesium 133.

As we discuss in Chapter 9, atoms absorb and emit photons of characteristic frequencies, and caesium 133 is no different. As a result, the second is defined as the duration of 9,192,631,770 frequency periods of the radiation corresponding to the transition between two energy levels of the caesium 133 atom.

The kelvin

Although in everyday life temperatures are expressed in Centigrade, Celsius, or Fahrenheit, scientists prefer to use a unit called the kelvin or the symbol K.

The BIPM define the kelvin as the fraction 1/273.16 of the temperature at which water can coexist as ice, water, and water vapour at a given pressure.

The temperature of zero kelvin (0 K) is often known as absolute zero. The cosmic microwave background has a temperature of 2.73 kelvin, very close to absolute zero. However, it's actually impossible to reach a temperature of absolute zero in practice because the object would always warm up to the temperature of its surroundings.

Non-SI Units

SI units are very convenient for allowing scientists who use different languages to share information. However, in the grand scale of the universe, they can become rather unwieldy because of the huge numbers involved. As a result, astronomers and cosmologists often use a number of special units to measure angles, lengths, and energy.

Angles

You probably remember from your school days that a circle has 360 degrees in it. For most everyday measurements of angles, degrees are sufficient. Thus, right angles have 90 degrees and governmental policies suffer from 180-degree reversals.

Angles are very important in astronomy and cosmology. This is because everything outside the Earth's solar system is so far away that you can for most purposes consider the stars and galaxies to be located on some giant invisible globe called the *celestial sphere*.

Astronomers describe the location of stars on that sphere using an angle and a time. The angle is called *declination*, which is the angle above or below the celestial equator (which corresponds to the Earth's equator but out on the celestial sphere). The time is known as *right ascension* and measures the position along the celestial equator from a point known as the first point of Aries – the point at which the Sun crosses the celestial equator on the day of the spring equinox in March. Right ascension is measured in hours, minutes, and seconds.

Right ascension is very helpful when looking through a telescope because it can tell you how long you have before a star disappears from view because of the rotation of the Earth.

As well as describing locations, angles are also used to determine sizes and distances. The Moon and the Sun, for example, both have an angular diameter of half a degree – because this is the same number, we experience total eclipses.

Distances also require the use of angles. The parallax method (see Chapter 5) of measuring the distance to a star requires the use of very small angles.

Because astronomers often use very small angles in their calculations, they subdivide the degree. Rather than using decimal fractions of a degree, astronomers subdivide the degree into 60. Each of these subdivisions is called a *minute*. Yes, that's right, the same as the 60th subdivision of an hour of time. All very confusing. To reduce the confusion a little, angular minutes are often referred to as *minutes of arc* or *arc-minutes* and have the symbol '.

These arc-minutes are still not small enough for some measurements and so astronomers divide them into 60 again. And guess what they call them – yes, *seconds*; or, to avoid confusion, seconds of arc, arc-seconds, or the symbol ".

So you can say that a circle contains 360 degrees, or 21,600 arc-minutes, or 1,296,000 arc-seconds.

Length

The scale of the universe means that the metre is not a very practical unit of measurement for some things. The following are a number of other measurements that astronomers and cosmologists use.

The astronomical unit

The *astronomical unit* (AU) is defined as the average distance from the Earth to the Sun, some 150 million kilometres. The AU is useful for measuring distances within the Earth's solar system. The orbit of Jupiter has an average radius of 5.2 AU, for example.

The light year

Outside the Earth's solar system, you must travel a very long way to the next nearest point of interest. The next nearest star after the Sun, Proxima Centauri, is a whopping 40 trillion kilometres away. Scientists don't want to have to write out that many zeros every time they do a sum, so another unit of distance comes in handy: the *light year*.

The light year, the distance that light travels in a year, is a good choice. In SI units, the speed of light is 299,792,458 metres per second and there are 31,556,926 seconds in a year. Multiplying the two together shows that light travels 9,460,528,412,464,108 metres in a year or around 9.5 trillion kilometres. Therefore, Proxima Centauri is just over four light years away from the Earth.

The light year is handy in another way. As we describe in this book, when you look out at the universe, you aren't seeing the universe as it is now but rather at some point in the past because light from distant objects takes some time to reach the Earth. But how long? Well, because the light year is by definition the amount of time that light travels in a year, the time the light has taken to travel to the Earth is the same as the distance in light years. So when scientists point a telescope at Proxima Centauri, the light that they see was emitted from the surface of the star just over four years ago.

The parsec

The parsec may be a familiar bit of jargon from science fiction films, but do you know what it actually means? The name is a contraction of parallax second and is related to the parallax method of determining the distance to a star (as discussed in Chapter 5).

The parallax of a star (the angle over which it appears to move in the sky between two observations spaced six months apart) can be used to calculate its distance from the Earth. One parsec is therefore the distance to a star that has a parallax of 1 arc-second (see the earlier section 'Angles').

How far is that? Well, it's 3.26 light years or about 31 trillion kilometres. That means Proxima Centauri is about 1.3 parsecs away from the Earth.

The electronvolt

Read any scientific paper on particle physics and you soon come across the electronvolt (or eV for short). This unit of energy is particularly well suited for use in the field of particle physics. Electric charges need energy to move in electrical fields and the electronvolt represents the amount of energy needed by an electron to move through a voltage difference of 1 volt. An electronvolt is equal to 1.602×10^{-19} joules.

Particle physicists have another use for the electronvolt – they use it to measure mass. How's that? Well, we know from Einstein that mass and energy are interchangeable through the equation $E=mc^2$ which we can rewrite as $m=E/c^2$.

This means we can express a mass in eV/c^2. An electron, for example, has a mass of 0.511 MeV/c^2 (well, when it's not moving it does). Using MeV (mega-electronvolts) or TeV (tera-electronvolts) is convenient because in a particle accelerator where you know the energy at which the particles smash into each other, you can get an idea of the mass of particles this is likely to create.

Key Equations

The following are the key equations that you need to know if you're to become an amateur cosmologist.

Kepler's third law

In Chapter 3, we explain Kepler's third law, which relates the length of a planet's year to the size of its orbit. In mathematical terms, you can write this as:

$$P^2 = ka^3$$

where P is the length of the planet's year and a is a distance called the semi-major axis. Because planets travel in ellipses (see Chapter 3), the diameter of the orbit is not the same in every direction. The *semi-major axis* is half the diameter of the orbit at the widest point of the ellipse.

Kepler's big eureka moment was realising that the k in this equation had the same value for every planet in the solar system. He didn't know why, but Newton eventually explained the phenomenon with his universal law of gravitation.

Newton's law of universal gravitation

Newton's law of universal gravitation equation shows the size of the gravitational force acting between two objects with masses m_1 and m_2 that are a distance r apart:

$$F = \frac{Gm_1 m_2}{r^2}$$

where F is the force and G is a constant equal to 6.67×10^{-11} m^3 kg^{-1} s^{-2} known as the *gravitational constant.*

This equation shows that gravity operates according to an *inverse square law* (move two objects twice as far apart and the force between them is a quarter what it was before). The equation also shows that the two objects in question exert the same force on each other (interchange the two masses and the force is the same). Hence the Sun and the Earth both exert the same force on each other as they orbit around their common centre of gravity.

The most famous equation – ever

Einstein's $E = mc^2$ is without a doubt the best known equation ever, apart from perhaps $1 + 1 = 2$, but what does $E = mc^2$ really mean?

The E stands for energy, m the mass of an object (at rest), and c the speed of light in a vacuum. Be sure to note the 'at rest' bit. Einstein also noted that an object's mass increases as its speed increases and this increasing mass is what stops anything travelling faster than the speed of light.

To see $E = mc^2$ in action, consider an electron, which has a mass of 9.1×10^{-31} kg when at rest. If you plug this number into Einstein's equation, you get 8.2×10^{-14} kg m^2 s^{-2}.

This unwieldy scientific unit is usually referred to by another name, the *joule* (symbol J). A joule is not very much energy when you consider that boiling a kettle takes many thousands of joules.

What the equation means is that if the electron were to transform itself into energy, that's how much energy you would obtain.

Hubble's Law

Hubble's Law is the equation that kicked off modern cosmology.

$$v = H_0 D$$

where v is the radial velocity of a distant galaxy, D is the distance to the galaxy, and H_0 is Hubble's constant, currently thought to equal 71 kilometres per second per megaparsec.

This equation shows that the farthest galaxies away from the Earth are travelling fastest and that space itself is expanding. This equation also points to a time in the distant past when all the galaxies would have been on top of each other – the point at which the Big Bang took place (see Chapter 6).

Index

• Symbols •

• A •

• B •

• E •

• H •

• *M* •

• P •

• S •

• T •

• U •

• W •

• Z •

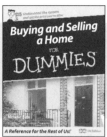

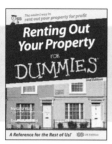

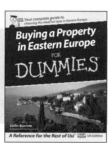

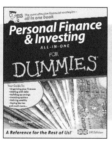

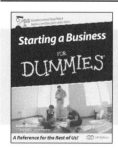

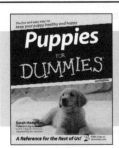